Gender Reboot

Eleonore Fournier-Tombs

# Gender Reboot

## Reprogramming Gender Rights in the Age of AI

Eleonore Fournier-Tombs
United Nations University
New York, NY, USA

ISBN 978-3-031-41389-6     ISBN 978-3-031-41390-2  (eBook)
https://doi.org/10.1007/978-3-031-41390-2

© The Editor(s) (if applicable) and The Author(s), under exclusive license to Springer Nature Switzerland AG 2023

This work is subject to copyright. All rights are solely and exclusively licensed by the Publisher, whether the whole or part of the material is concerned, specifically the rights of translation, reprinting, reuse of illustrations, recitation, broadcasting, reproduction on microfilms or in any other physical way, and transmission or information storage and retrieval, electronic adaptation, computer software, or by similar or dissimilar methodology now known or hereafter developed.
The use of general descriptive names, registered names, trademarks, service marks, etc. in this publication does not imply, even in the absence of a specific statement, that such names are exempt from the relevant protective laws and regulations and therefore free for general use.
The publisher, the authors, and the editors are safe to assume that the advice and information in this book are believed to be true and accurate at the date of publication. Neither the publisher nor the authors or the editors give a warranty, expressed or implied, with respect to the material contained herein or for any errors or omissions that may have been made. The publisher remains neutral with regard to jurisdictional claims in published maps and institutional affiliations.

This Palgrave Macmillan imprint is published by the registered company Springer Nature Switzerland AG
The registered company address is: Gewerbestrasse 11, 6330 Cham, Switzerland

Paper in this product is recyclable.

*This book is dedicated to my two sons, Julien and Alessio.*

# Foreword

We need a gender reboot! A restart, an upgrade, a systems change. And it all starts with a new way of articulating how all our systems are intricately interconnected, and particularly how analog systems are being permanently wired into our new digital universe with AI.

We can't genuinely think about how the digital world can bring about a new and more equitable world until we understand and combat how historical social and gender norms are inadvertently being embedded into AI and new emerging technology. And we can't begin this process of understanding until we openly recognize the largely unseen and gendered systems operating at gale force around us. Care work is a prime example of this.

If COVID 19 and its shutdowns taught us anything it is that our systems are cracked, functioning on the backs of our care workers—from the (mostly unpaid) home to the (vastly underpaid) hospital. Until we fix the notions and reflexive rituals of who does care work, and how it is valued; what strength is and who can be strong (I loved the Gender Reboot chapter on strength training and stereotypes); who a leader is, who can lead, and what new leadership could look like; how the educational system reinforces stereotypical directions (from coloring books to time on the soccer pitch); to the actual data we are using to make a funhouse mirror digital world—we will have an uphill, even Sisyphean climb to gender equality.

Gender is one of those concepts and realities that you cannot unsee once you have seen it. And then, once you have seen it exists, you see it everywhere in the structures and myths and assumptions that surround you in daily life. It operates wherever you live in the world, at whatever economic strata, race, class or caste, at whichever intersections. And indeed the more intersections in your lived reality the more and more gender plays a brutal part in the rules you are told to, or must, live by.

This is what Gender Reboot and Eleonore Fournier-Tombs explore in a most powerful and accessible way. Unless we understand the systems surrounding us, then pause/reflect/define what we value, we will be doomed to inject the old gendered values so slowly being stripped from the analog world directly into our new and powerful digital systems. And the velocity and scale at which AI is being adopted will also exponentially increase the ways we will be expected to perform our gender, (as gender and gender roles are nothing but performative).

So this is a critical moment to look at how gender is performed all around us and how we perform it. It is a moment to change the unwritten rules that constrain. It is a moment to ensure that we do not replicate the confinement of gender, (made more apparent during our collective COVID confinement), and embed it into our AI digital universe.

Gender Reboot takes these concepts and makes them sublimely accessible using interviews, hard facts, and personal experience then weaves a story of how critically we all need to be aware of the gendered systems functioning almost invisibly around us—systems that create the structures that we cannot emerge from, or change, until we are aware that they even exist.

Read and think about how you want to take part in the grand Gender Reboot. Then reach out and join in as we collectively reflect, reconstruct and produce systems and technologies for a more gender-just, more equitable, kinder and more creative world that works for us all and brings everyone along for a wonder-full reboot and ride.

Geneva, Switzerland  Caitlin Kraft-Buchman
2023  Women at the Table

# Preface

This book aims to raise awareness about the multidimensionality of women's rights and gender roles in the age of AI. It is not exceedingly academic, and aims often at a more conversational style, drawing not only from my research, but also from personal experience and intuitions. There is something extremely intimate about gender—how we perceive ourselves, our roles in our private lives, our bodies, our work. I therefore wanted to try to find a bit of this intimacy while also covering what I think is important ground.

Each chapter therefore aims to present a different dimension of current gender norms, which together can be seen as painting a picture of AI and the feminine. At the beginning of each chapter, I present a hypothetical scenario, always drawn on real news items or research, which explains how a certain way of stereotyping, excluding, or discriminating women in AI would impact an individual woman's life. These examples resonate with me, and they may resonate with you too. At the end of each chapter, I also pose certain questions. This is certainly the professor in me, but I hoped to shape them so that a class in society and technology might use them to stimulate an interesting discussion.

But this book is not only for students, it's also for policymakers and AI developers. As we set regulations and guardrails for AI tools to be used safely, I hope to inform a more holistic, systemic conversation about women's rights. I want to help us get out of being reactionary when it comes to suddenly noticing a bias or a harm, and thinking more carefully about what these new tools might mean for the place of women and men in society.

So, in these pages, you will not only read about AI—you will read about the history of women in the workforce, about leadership, about sports, about education. Most importantly, you will read about the roles that women have had, and are striving for, and what this means in relation to the roles that men might want too. Today, we are in the age of AI, and tomorrow, we will be in the age of something else. We may still be debating gender roles and women's rights, but I hope we will have come a little bit further in allowing all genders a freedom to participate, lead, and self-actualize.

New York, USA                                        Eleonore Fournier-Tombs

# Acknowledgments

This book took over three years to write, from the first iterations in March 2020 to being able to finally send the manuscript to Palgrave Macmillan in March 2023. In the process, I benefited from the invaluable help of many colleagues, friends, and family members. First, I would like to thank my two research assistants—Christina Darvasi and Brittany Guillory, both successful professionals, who took time from their busy schedules in Fall 2020 to conduct research on women's rights. My sister, Myriam Fournier-Tombs, provided unconditional support and reviewed many chapter drafts for me, as I tried to decide whether this book would be published in French or in English. My other siblings, Nate and Iona Fournier-Tombs, who always encouraged this project and were willing to brainstorm ideas. My girlfriends who inspired me, re-read drafts and articles, and gave me their time in a multitude of ways—Marie Glorieux, Veronica Bagdoo, Dagmar Hertova, Laurence Mathieu-Léger, Alexandra Hall, Melissa Offner, Subhra Bhattacharjee, and many more amazing women whose perspectives imbue the pages of this book.

I'd also like to acknowledge the University of Ottawa Faculty of Law and United Nations University, my two main academic homes while

I wrote this book. Both venues provided me with valuable opportunities to dive deeper into gender and AI, whether it be by giving me venues to write chapters and give webinars, or by facilitating applied research projects with the International Telecommunications Union and UN Women. In this context, I particularly want to thank Céline Castets-Renard, who has been a mentor and an immense support during this period.

I am immensely grateful for the support of my coach, Robin Hornstein, who was the first person (outside of friends and family) to hear about my book-writing dreams. I also want to acknowledge Sonia Sauvette and Lison Lescarbeau, who provided editorial support to some of my earlier work on gender norms in French, much of which made its way into this book.

And finally, I must thank my husband, Curtis Hendricks, for his unwavering support and patience. He heard talk about this book (the lows, the highs, the doubts, and the breakthroughs) nearly every day for three years and still found somewhere the energy to review drafts and discuss ideas.

# Contents

| | | |
|---|---|---|
| **1** | **Introduction: Reprogramming Gender Norms** | **1** |
| 1.1 | Reprogramming Gender Norms—A Trajectory | 4 |
| 1.2 | Salsa Dancing into Equality | 5 |
| 1.3 | Why this Book, Now? | 7 |
| 1.4 | The Subconscious Propagation of Biases | 8 |
| 1.5 | Giving Women Back Their Place | 10 |
| 1.6 | The Ethics of the Future | 12 |
| 1.7 | Generating Returns Before Building Meaning | 12 |
| 1.8 | Ethics and Policy of AI | 13 |
| 1.9 | The Public and the Private Sphere | 15 |
| 1.10 | Can We Reprogram Gender Norms? | 16 |
| References | | 17 |

**Part I Work**

| | | |
|---|---|---|
| **2** | **Woman Is to Housewife as Man Is to Programmer** | **23** |
| 2.1 | The Right to Work | 24 |
| 2.2 | Stifled by Stereotypes | 28 |
| 2.3 | COVID-19 and the Second Shift | 31 |

|        |                                              |    |
|--------|----------------------------------------------|----|
| 2.4    | The Family or Career Paradox                 | 34 |
| 2.5    | Multilateral Efforts against Discrimination  | 35 |
| 2.6    | Reprogramming Gender Rights at Work          | 37 |
| 2.7    | Suggested Discussion Questions               | 38 |
| References |                                          | 40 |

## 3 Rebalancing the Distribution of Unpaid Duties — 43
- 3.1 First Generation Househusbands — 46
  - 3.1.1 Can a Midwife Be a… Man? — 50
  - 3.1.2 Fatherhood and Returning to the Roots — 52
- 3.2 Lean In or Lean Out? — 54
- 3.3 Learning to Parent — 57
- 3.4 Robot Teachers — 59
- 3.5 Corporate Policies and Social Norms — 60
- 3.6 Suggested Discussion Questions — 63

References — 63

## 4 Leadership in the Public Sphere — 67
- 4.1 Barriers to Women's Leadership at Work — 68
- 4.2 The Trap of Long Work Hours — 70
- 4.3 Negative Perceptions of Female Leaders — 72
- 4.4 The Third Way: Consulting — 73
- 4.5 Women and Motherhood in Government — 76
- 4.6 Suggested Discussion Questions — 79

References — 80

## Part II  Being

## 5 Strength and the Power of Sport — 85
- 5.1 Women in Strength — 87
- 5.2 The Superwoman Schema — 90
- 5.3 The Sports Pay Gap — 91
- 5.4 Sport for Equality — 92
- 5.5 Empowerment in the Gym — 94
- 5.6 Suggested Discussion Questions — 96

References — 96

## 6    Physical and Virtual Safety                                99
6.1    Types of AI Used in Gender-Based Violence               102
6.2    The Tay Twitter Bot                                     103
6.3    How Media Can Cause Violence                            103
6.4    AI to Make Women Safer                                  108
6.5    Suggested Discussion Questions                          111
References                                                     112

## 7    Sexual Stereotypes and Body Image                       115
7.1    Physical Stereotypes and AI                             116
7.2    Robots that Never Age                                   119
7.3    Fembots and Sexbots                                     120
7.4    From Physical to Digital Representations                122
7.5    Protecting Women from Sexual Violence                   124
7.6    Suggested Discussion Questions                          126
References                                                     127

## Part III  Learning

## 8    Learning Gender Roles                                   131
8.1    Gender Education: A Global Challenge                    132
8.2    Male Role Models                                        134
8.3    Preparation for the Workforce                           135
8.4    Gender Neutrality in Education                          138
    8.4.1   Radical Solutions in Iceland                    139
8.5    Gender Compensatory Activities                          142
8.6    Suggested Discussion Questions                          144
References                                                     145

## 9    Machine Learning and Collective Unintelligence          147
9.1    How Recommendation Systems Learn                        148
9.2    Passing the Turing Test                                 150
9.3    Scrum and Financial Incentives                          153
9.4    Collective Unintelligence                               155
9.5    Suggested Discussion Questions                          156
References                                                     157

| 10 | Conclusion: Gender Reboot | 159 |
|---|---|---|
| | References | 165 |

**Index** 167

# List of Tables

| | | |
|---|---|---|
| Table 2.1 | Historical obstacles to the participation of women in the labor market | 26 |
| Table 3.1 | Translating non-gendered pronouns using AI | 45 |
| Table 4.1 | Current state of women in leadership positions in Canada, the United States and Mexico | 68 |
| Table 4.2 | Average hours worked and GDP | 71 |
| Table 6.1 | Types of threats to personal security | 101 |
| Table 6.2 | AI systems used for information production and dissemination | 102 |
| Table 8.1 | Hjalli approach curriculum | 141 |
| Table 9.1 | Sources of training data for GPT-3 | 152 |

# 1

# Introduction: Reprogramming Gender Norms

In early March 2020, my son had a quiet spring break. He was a member of the "Pre-K 2" class at his small primary school. "That's us," he told me proudly as he showed me his class photo, which he had stuck with way too much tape to our fridge. We were watching the news about a new virus that seemed alarming, yet so far away. Then, suddenly, it was a global crisis. The school closed for a few days. Then a week, then quickly, six weeks. My husband, the calm one, tried to reassure me: "There is no way that they will cancel the school year." I wasn't so confident. "He's not going back," I thought, looking up at the photo.

For the next few weeks, we settled into a routine, trying to work together while looking after our boy. I would supervise him in the mornings, when he was doing math and learning his letters, and my husband would take him in the afternoons, to play hockey and draw. There was also a dedicated time—two hours after lunch, when we went on a "ninja" adventure, which consisted of running and jumping in the park by the river. After these sessions, I would hurry back to my video calls, a bit disoriented, juggling it all as poorly as you might imagine.

When schools around the world began to close, I realized how dependent we had been on this institution. It was a level of stress and

© The Author(s), under exclusive license to Springer Nature
Switzerland AG 2023
E. Fournier-Tombs, *Gender Reboot*,
https://doi.org/10.1007/978-3-031-41390-2_1

exhaustion that I had not experienced before. Finding it extremely difficult to concentrate during the day, I worked almost every evening until the early hours of the next morning to catch up. I started speaking to other women—my friends at first and then gradually other young professional mothers. Almost all of them were doing most of the childcare at home. These women were professionals—researchers, project managers, journalists, and often also either the main breadwinner of the family or an equal contributor to the household income.

I was struck by the fact that even in situations where the woman earned more than her partner, she now had to juggle most of the childcare duties at home. Women everywhere were losing sleep, compromising their work, their health, feeling stretched. One of my friends was juggling three children under the age of six, mostly on her own, and working nights on large contracts. She had always been one of those super-human women who does everything and still has time to work out every day, and yet I could see that she was completely at her wits' end.

There were stories, though, in which the father took the leading role. In one case, the husband was laid off in March, at the beginning of the shutdown. His wife, who was pregnant, continued to work remotely for a university. He took care of their four-year-old daughter during working hours, organizing impressive cross-training circuits for her and taking her to the park.

As I spoke to more and more people, I realized that fathers that took a truly equal part of the care of their children, let alone a larger part, were often still an exception to the norm. But women, by juggling children, housework, and work performance, were still doing a second full shift, this time without the invaluable support of school. Digging around for answers, I was struck by a sentence from Statistics Canada's time use survey.

"...*there is evidence that men have not increased their participation in unpaid work to the same extent that women have increased their participation in paid work, or that women have decreased their participation in unpaid work.*" That is, in the last few decades, women have started to take more and more responsibility at work without reducing their domestic workload.

This surprised me. Why haven't they gotten a break at home? If this is true, it's no wonder that women are underrepresented on boards, in senior management and in just about every position representing wealth and power—they are exhausted!

At the same time, as a data scientist, I was working as a researcher on the social impacts of new technologies, especially artificial intelligence. I began to be interested in the dynamics of technology and gender, notably women's rights. While women were struggling, during the pandemic, to balance work responsibilities with extremely challenging care demands, articles were coming out about how algorithms discriminated against women. Amazon, for example, had been found to use a resumé-sorting algorithm that discriminated against candidates who included "feminine activities," such as being the president of the women's rugby team or having attended the traditionally female American university Barnard College. UNESCO had also just come out with a report on sexist stereotypes in AI, arguing strikingly that voice assistants such as Apple's Siri normalized sexual harassment and violence against women by presenting a flirty and submissive female persona.

Sexist machines, however, exist in a context in which women have been discriminated against in nearly all domains of public life for as long as any of us can remember. Algorithms are trained by male-dominated teams on gender-biased data, and quality tested on largely male subjects. This is not new to artificial intelligence, as, for example, women have a higher risk of dying in car accidents because seat belts and air bags are tested on male-proportioned dummies. Advances in women's health technologies have also long lagged men's, as a striking number of women suffer from post-partum pelvic floor injuries, endometriosis, and difficult menopause with very little treatment options.

The use of all these technologies, however, has been accelerated by the pandemic, which has seen people work remotely and virtually more than ever before. Children who had Internet-connecting devices went to school online, adults worked on their laptops from their homes, engaged in interminable Zoom calls, and people ordered food and other items from the Internet.

This made me wonder—if technologies could discriminate against women, could the opposite also be true? That is, would it be possible to reverse gender biases in artificial intelligence and other technologies so that they could better serve women? Could these changes be so impactful that inequalities in other areas, such as work, childcare, and leadership, would be also reduced?

So began the work for this book, which represents several years of research and experiments in technology and gender, interviews with dozens of women and men in Canada, the United States, and Mexico, and much brainstorming, to understand the way in which reversing gender biases in technology might also reverse gender biases throughout society.

## 1.1 Reprogramming Gender Norms—A Trajectory

The objective of this book is to take you, the reader, on a journey through the many facets of gender biases, from work to leadership to bodies and childcare. Throughout these themes, I will examine how technology, and especially artificial intelligence, has perpetuated these biases. I will then explore what we could do differently when designing them, whether more balanced data, more representative teams, or better policies, to reverse not only the biases in the technologies themselves, but also in society.

As a data scientist, I have worked on artificial intelligence algorithms, notably to implement *predictive analytics*, which is that branch of AI which allows us to project trends in past data into the future. During the pandemic, I became intimately familiar with the impact of this field as I worked on predicting the spread and growth of the COVID-19 virus in countries experiencing other humanitarian crises. This experience allowed me to better understand how biases could insert themselves in the algorithms and what we might be able to do to prevent that. I also realized how mysterious artificial intelligence seems to those outside the field, a mystique which allows certain people to get away with unethical practices by claiming that it is all a black box. I don't believe that. I've always understood how my algorithms work and I think that in most cases, other data scientists do as well. And this mean that we are not powerless in the face of a technology we cannot control. Rather, we can use this powerful tool to help us win the battle for gender equality.

The book is therefore structured as follows. After this chapter, the first section—Chapters 2, 3, and 4—will deal with women and work. Chapter 2 will discuss women's right to work and representation in different professions; Chapter 3 will discuss care activities (such as childcare and the care of elders) and unpaid work; and Chapter 4 will discuss leadership and power in the public sphere.

Then, we will move on to the second section. Chapters 5, 6, and 7 will deal with women's bodies. Chapter 5 will address physical stereotypes, notably in sports and strength; Chapter 6 will discuss beauty and sexualization of women; and Chapter 7 will discuss women, peace, and security.

Finally, the third section—Chapters 8 and 9—will respectively discuss early childhood education and paths to STEM (science, technology, engineering, and math) education.

Each of these chapters will contain a description of the challenges facing women in these domains, the impact of artificial intelligence and other new technologies on these challenges, and some clues as to how to reprogram gender norms. The book will conclude with a summary of all policy recommendations discussed throughout the book, as well as the path forward for research, technological investment, and new regulations.

## 1.2 Salsa Dancing into Equality

My favorite research methods text is "Salsa dancing into the Social Sciences" by Kristin Luker. In this book, Luker suggests combining quantitative and qualitative methods to get a more complete picture of the issue being researched.

Like Luker, I have employed several methods in this book in order to reflect the complexity and multidimensional aspects of the gender equality issue. I will present as diverse a view as possible. The researchers I have worked with contribute to the development of these multiple perspectives. With the help of two research assistants, Christina Darvasi and Brittany Guillory, based in Mexico City and New York, respectively, I was able to speak with a variety of people: women and men; some related, some not; some cisgender, some LGBTQ; some black, some white or Hispanic. Their generosity in sharing their stories with me enriches these pages and helps to elaborate a plural reality.

I have included several of the interviews in this book. I have let the people speak for themselves as much as possible. Their surnames are often omitted to preserve their anonymity, unless they speak as professionals and are comfortable being quoted in full. A great deal of background research and statistics complements the discourse to allow for comparisons across countries.

The women and men interviewed are chosen according to specific criteria. I have selected mainly women with professional careers who feel that there is still a lot of work to be done in terms of gender equality. I spoke to many working mothers, who shared with me their perspective on the second shift at home, and the struggle to be heard and respected at work. I also spoke to women who were not mothers who were still under pressure, whether it was to give up their careers so that they could better focus on their families.

The men I spoke to were all, in one way or another, disillusioned with the traditional patriarchal system and eager to forge a new path for masculinity. One of these men became a youth advocate in his neighborhood. A very involved father, he posted on Facebook the activities he was doing with his children as a role model for other men. Another became a social worker and took paternity leave to spend time with his newborn baby and his partner. I also spoke to experts, who gave their views on gender norms and how to change them. Naima Beckles, for example, a doula from New York, describes how she trains men to move beyond their fears about childbirth and infancy.

Finally, in order to understand the themes discussed in the broader context of gender norms, I supplemented all interviews with statistical data and historical milestones. At first, I focused on three geographical areas: Quebec (within Canada), the United States, and Mexico. As I wrote the book, however, I had several opportunities to conduct research on women and AI in other countries too, and I have included several of these examples here.

## 1.3  Why this Book, Now?

One fall morning in 1960, a 26-year-old researcher by the name of Jane Goodall was observing the behavior of chimps in Gombe National Park in Tanzania. Squatting in her khaki shorts and tennis shoes, she was taking notes in her field book, watching one of the chimps she had gotten to know best, and who she had named David Greybeard. David Greybeard seemed to be poking a termite mound with blades of grass. Not wanting to startle him, she stayed at a distance until he had wandered away. As she approached the termite mound, she picked up a blade of grass and inserted it inside. Much to her surprise, a termite grabbed the grass with its jaws, allowing her to fish it out. A little later, she watched as David Greybeard stripped a twig of its bark to build a sturdier termite-fishing rod. Over time, she made more discoveries, observing chimps using rocks, twigs, and branches for activities such as drinking water, eating, cleaning themselves, and even fighting.

The discoveries rocked the research community. After all, technology was what made us human. If chimps were able to make tools, perhaps the boundaries between them and us were not so clearly defined as we once thought. Since then, we have observed other animals using tools. Orangutans, for example, make whistles out of leaves to ward off predators. Octopuses use shells for armor and camouflage, as was beautifully captured by the documentary film *My Octopus Teacher*.

As humans, we are by far the animal that is the most apt at transforming raw materials into tools for our own use. Stripping bark off of twigs is one thing, but working remotely, as I am doing now, with the Internet at my fingertips, an iPhone playing whale sounds (to help me focus), and a watch laid out beside me, is quite another. I am in a heated house, with electric light as well as a candle burning, a piano in my peripheral vision, and Christmas lights blinking on and off on an artificial tree. My desk is too messy, so I am at the kitchen table, which also has the salt and pepper grinders, as well as an assortment of crayons and markers. The wooden table was built with jointers and planers, the shelves were fastened to the wall using screws and a drill, everywhere I turn my head I can see the product of human ingenuity. Every element on my body—my clothes, the color in my hair, my glasses—every aspect of my daily routine—writing, driving with my children, going sledding, listening to music, cooking—is affected by tools that we created.

Humanity's collective fascination for technology is completely understandable. It is through technology that we learned how to make fire. It is through technology that we were able to create carts that would carry loads for us. Technology is what helped us create some of the most incredible wonders of the world, such as the Egyptian pyramids (although we still don't quite understand *how* those were built).

Technology has always changed what it means to be human. While technological change happened gradually, over several generations, over the last century, the rate of change has very much increased. Each recent generation has had a very different experience than the generation before—could our great-grandparents have expected to see man walk on the moon, self-driving cars, the arrival of the Internet and smart phones and deep learning? Perhaps the 1900 Paris Exposition showcased many technological innovations, such as elevators, moving sidewalks, audio-recording machines, and talking film. However, we were a long way then from the world we live in now, particularly when it comes to communications technologies, travel, and computation. In just a few decades, the framework within which we interact with technology has been completely upended.

## 1.4 The Subconscious Propagation of Biases

Technology is therefore a key concern of our society. We hear about the latest technological developments in the news every day. Technological ethics, once the domain of philosophers and doctors, is now a common consideration, especially in relation to artificial intelligence. Do algorithms have biases? How should they be created and used? How do we understand privacy? There is a lot of excitement about what these new technologies can do for us, but there are also many fears and concerns. After all, we do not have full control over these technologies and many of them present serious risks to society that need to be considered.

Timnit Gebru is the former co-director of AI-Ethics at Google in the United States, where she was fired in December 2020 for refusing to retract a research paper on the ethics of large language models. She is also the founder of the non-profit Black in AI and a strong advocate of ethics in artificial intelligence. In a conference, she describes the pervasiveness of artificial intelligence, which "is being used to manage our electric system, power systems, trading, trying to figure out who to hire, and who not to hire."

"Invisible algorithms," she says, are everywhere in our lives, and there are many unintended consequences that might come because of that omnipresence. For example, the United States government recently planned to partner with software companies to track migrants' activity on social media and evaluate whether or not the person would make a good immigrant. According to Gebru: "The current AI tools are not robust enough to make decisions in such high-stakes scenarios."

The problem that Gebru cites is related to the use of artificial intelligence algorithms to predict human behavior. This issue is incredibly philosophically complex. Algorithms have been used, for example, to predict recidivism—meaning how likely a criminal is to reoffend when he or she is released. The idea behind this is to use these algorithms to determine the lower-risk convicts, which could be released, and keep a closer watch on those that were deemed higher-risk, perhaps even postponing their release. As Virginia Eubanks wrote in *Automating Inequality*, these algorithms propagated the biases embedded in the United States' incarceration system, considering poor, black convicts to have a much higher recidivism risk as other convicts, regardless of their crimes.

Another major problem cited by Gebru relates to the impact of prediction errors, which are not always captured during the technology's testing period. An example she cites happened recently when a young Palestinian posted a message reading "Good Morning" in Arabic on Facebook. Facebook's translation algorithm mistakenly translated this to "Attack Them," and the young man was arrested by the Israeli authorities. She argues that the Israeli authorities did not use their critical judgment to verify the original language and context of the post, due to automation bias. "We are propagating biases in ways that we don't even understand right now," she says.

These examples show two of the main issues that technology ethicists are grappling with—bias propagation and prediction errors, both of which have the potential to increase inequality. In terms of gender norms, these issues are linked to two major considerations. First, there is the underrepresentation of women in technological work. Second, there is the lack of consideration of the impact of these technologies on women.

As artificial intelligence programmer Kriti Sharma says, chillingly: "How many decisions have been made about you by AI, and how many of those were based on your gender, your race, or your background."

## 1.5 Giving Women Back Their Place

Over the course of human history, women have led many technological advancements. Unfortunately, many of their contributions have been erased from history books and are therefore difficult to track. The movie *Hidden Figures*, for example, follows the lives of three African American mathematicians, all women, who play a critical role at NASA during the Cold War. Before this movie, I hadn't heard of many women in the past, being brilliant mathematicians. However, a little digging uncovers many important women in mathematics and technological development, as far back as Hypathia, a prominent philosopher, astronomer, and mathematician of the fourth century.

In her book: "Invisible Women: Exposing Data Bias in a World Designed for Men," Caroline Criado Perez writes that: "Most of recorded human history is one big data gap. Starting with the theory of Man the Hunter, the chroniclers of the past have left little space for women's role in the evolution of humanity, whether cultural or biological. When it comes to the lives of the other half of humanity, there is nothing but silence."

Reading history books still gives us the impression that women have not contributed much to the advancement of the world. This is a big problem, given that technological development is one of the most important achievements of humanity, as well as the source of power for societies throughout history. By this, I mean that throughout history, civilizations that have had a cultural and geopolitical impact have done so through the technologies they have developed. The major empires in human history, whether it was the Egyptians with their papyri, pyramids, and agricultural advances, the Romans with their roads, aqueducts, and surgical tools, the Incas with their astronomy and architecture, and so on—all went through significant and innovative achievements in the field of technology.

Yet, from what we are taught, we get the impression that there has not been a single important female doctor, astronomer, or architect in human history, before the last few decades. I, for one, have my doubts. In a presentation for INVDI Technologies, data scientist Samira Sadeghi writes: "Everyone knows that the great mathematicians in history were all men. But everyone is wrong." She goes on to mention several uncredited female leaders in the field of mathematics and technology, such as Theano, who led the school of Pythagoras after his death. She wrote many books on astronomy, mathematics, physics, and psychology, perhaps the most important of which is The Middle Way Theorem. It is estimated that she taught about 30 women in this school, in addition to the male students. And what happened to those 30 female mathematicians? Did they really go back home after their training to do their housework? In my opinion, we are missing a big part of the story—the contribution of women who have become invisible.

Some information about women's contributions to technology throughout history is available with a little digging on the Internet or in library archives. It is important to know that these contributions exist so that we can look for them. It is suspected that many of women's contributions are simply lost. The suppression of women's achievements has accelerated their exclusion, making us believe that they have been spectators of humanity's discoveries. This is why the current women's movement in artificial intelligence is so important. This field, which will bring about a major upheaval in our lives, cannot develop by excluding the half of the population it will impact.

## 1.6 The Ethics of the Future

Bruno Latour, a French philosopher of science and technology, is one of many theorists who have argued against the concept of object neutrality. Objects, he argues, are always embedded in the social norms of the society that developed them. In one of his many articles on the subject, entitled "Technology is a society made sustainable," he argues that to understand power and domination in society, researchers should turn to the study of physical objects: "Society and technology are not two ontologically distinct entities, but rather phases of the same essential action," he writes.

This way of conceptualizing allows us to understand that technology merely reflects the values already present in our society. If we adopt Latour's approach, then the point is not so much that technology is biased, but that our society is.

This view, which I share, contradicts some of the conventional wisdom that technology is always neutral. It is difficult to accept that the models and objects of artificial intelligence are sexist simply because our society is sexist. In the same way that we believe technology can solve all our problems, we want to blame it entirely when something goes wrong.

Several other science and technology researchers have echoed these discomforts. Diane Martin, for example, argues that "technology embodies the values of developers." She argues that ethics should be taught as a priority in computer science curricula, as "it is unethical to ignore the values embedded in technological artefacts." Although this is changing rapidly, computer scientists and engineers are still not sufficiently trained in the ethics of technology.

## 1.7 Generating Returns Before Building Meaning

Many technologies have had a negative effect on women. Sometimes they have been corrected, as was the case with car seat belts. To take an example from a different sector, the same problem arose in the pharmaceutical industry, where there was no requirement that drugs be tested on women or other more minority groups in our societies. Medications have been shown to have adverse side effects for groups that had not previously been considered in testing. Little by little, this industry has adjusted to the diversity of the population.

We live in a capitalist economy that pushes companies to make high returns for investors rather than producing value for communities. Computer scientists are no exception to this rule. They are under great pressure to create accurate and efficient algorithms to meet financial goals. A classic way to test the accuracy of a model is to use a dataset whose predicted value is already known. For example, let's say we are trying to predict whether a face is female or male. The simplest way to do this is to take a database of photos that are already labeled as "female" or "male" and use part of the database to create a model based on these images. This model is then used to predict the sex of the rest of the images, the result of which is compared to the actual sex. Often in machine learning this is done by using 80% of the dataset to build the model and 20% of the dataset to test it. Once the model can predict the sex of the images in the test dataset most of the time (i.e., 95% of the time), it is made available to the public.

Often the model encounters data on which it has not been trained or tested. For example, many facial recognition systems were originally trained on the faces of the developers who built them—young white men. When they were used to recognize the faces of people of color or women, they made gross errors. Over time, these models might be updated to be more inclusive, but only if the companies that build them recognize the value of this improvement. The promotion of social and equality dimensions is only collateral benefits of economic value improvement projects. Encouraging diversity in representation is not part of the mandate.

## 1.8  Ethics and Policy of AI

Today, there are many ethical standards for technology, particularly in relation to artificial intelligence, but still few regulations. UNESCO has spearheaded a recommendation on the ethics of artificial intelligence, which provides normative guidelines for artificial intelligence, data, and digital technology. Such a multilateral document is very important, but only the first step to address gender inequalities. In the coming years, much stronger laws at the national level are needed that protect women from biases in artificial intelligence.

In April 2021, the European Commission published a draft law on AI that proposes a method to regulate artificial intelligence products before they enter the market. This draft law introduces the notion of prohibited technologies and high-risk technologies. High-risk technologies—such as AI used in police surveillance, biometrics applied to migrants, AI in schools, in human resources systems or to determine eligibility for a bank loan—would have to obtain safety certification, like the process a drug goes through before it is approved for the market. Prohibited uses of AI include social scoring, subliminal manipulation, and real-time, remote biometric identification systems. The document mentions the possible biases of this technology, and it will be up to the developer of the technology to prove that there is no gender bias.

For gender and technology, this is an important start, not least because the document begins to acknowledge gender bias in artificial intelligence and presents some solutions. For example, in the introduction, it talks about recruitment technologies, explaining: "Throughout the recruitment process and in the evaluation, promotion or retention of people in work-related contractual relationships, these systems can perpetuate historical patterns of discrimination, for example against women [...]." For a text of about 100 pages, this is not a very lengthy mention, but it still means an important step in the regulation of AI globally.

Other countries, such as Canada, have also introduced bills that begin to touch on gender bias in artificial intelligence. In Quebec, the Commission des droits de la personne et des droits de la jeunesse (CDPDJ) has begun to assess gender discrimination in technology, explaining in its June 2020 brief that gender discrimination can be the subject of complaints, particularly when it affects employment and services.

If we take an optimistic view, we can believe that the trend is beginning to change, and that gender biases in technology will begin to diminish. However, these biases exist in a much larger context, and it is unlikely that they will change independently of the inequality in our society. Technology can also be an important vehicle for spreading egalitarian gender norms and promoting human rights globally. It will depend on the rules and ethics that are followed.

## 1.9 The Public and the Private Sphere

In political theory, society is often separated into two areas of study—the public and the private sphere. In the private sphere, we find everything that is related to personal life and the household, such as family, the home, and intimate relationships. In the public sphere, we find everything else—work, political participation, interactions with those outside of one's inner circle. The terms were attributed originally to Jürgen Habermas, a German sociologist and philosopher who was born in 1929.

Habermas argued that the public sphere was "*made up of private people gathered together as a public and articulating the needs of society with the state.*" Conversely, Mary Ann Tétreault, an American Political Scientist who died in 2015, wrote that "private space at its most fundamental is where a person can *be* without being *seen*." She found that the association of women with the private, the unseen, the personal, and men with the public, the societal, and the influential, has roots in Ancient Greece, where only free adult males were considered citizens. Unlike women, children, and slaves, a "citizen was an autonomous agent who came together with others like himself to create a good society." And while citizens moved freely between the private and the public spheres, women operated only in the private one.

Most people would agree that this way of thinking does not reflect what we have tried to achieve in modern democracies, specifically as was codified in the Universal Declaration of Human Rights, which begins with the statement: "All human beings are born free and equal in dignity and rights." It goes on to detail, in its 30 articles, how every human being should have equal opportunity and influence, in all areas of society. And yet, when we look at the struggles of feminism in recent years, it's difficult not to think that we are stuck in ancient and defunct paradigms.

In fact, much like the women in Ancient Greece, women today are still responsible for the private sphere and are immensely pressured to prioritize it. Although in most societies (but unfortunately not all), they are now free to move between the two, the responsibility that they have in the private one is such that they may not have time for the public sphere. Moreover, women still face enormous amounts of discrimination in the public sphere, whether it is bias in electing political representatives, belief that women don't make as competent leaders as men, or the lack of women at the most influential echelon in most sectors.

## 1.10 Can We Reprogram Gender Norms?

Whether or not we can, in fact, reprogram gender norms is a thorny question. After all, some might say, if the problem is societal, rather than technological, shouldn't we just fix society first, and technology will follow? This does not, however, account for the vicious cycle of gender and technology, where algorithmic biases reinforce discrimination and negative stereotypes, which in turn cause more biased algorithms… possibly forever. Take the example of biased human resources algorithms in big technology companies. These companies already have a disproportionately small number of women developers working there. That is part of the reason for which the algorithm was not properly quality tested. If the algorithms weed out female applicants at a greater rate than male applicants, the number of women working there is likely to get smaller. This will lead to more biased data showing that men are more likely to be hired and less women to bring a female perspective when sexist algorithms are produced. If, on the other hand, a human resources algorithm is developed which accounts for qualities and qualifications that are more common in women, then more women are likely to be hired. Men, too, who tend to fall through the cracks because they might have a background different to the one modeled, now have a better chance at working in these companies and developing better algorithms. These algorithms can go on to affect what content people are shown in streaming videos, how likely someone is to get a home loan, and much more.

So, it's not just that we *can* reprogram gender norms, it's that we *must*. Given the acceleration of gender inequality and the increase in the use of artificial intelligence, both caused by the pandemic, this is the moment to address head on one of the most pervasive and long-lasting injustices in our society.

## References

Buolamwini, J., & Gebru, T. (2018). *Gender shades: Intersectional accuracy disparities in commercial gender classification*. Paper presented at the Conference on fairness, accountability and transparency.
CDPDJ. (2020). *Mémoire à la commission d'accès à l'information sur le document de consultation: Intélligence artificielle*. https://cdpdj.qc.ca/storage/app/media/publications/memoire_consultation_CAI_IA.pdf
Dastin, J. (2018). Amazon scraps secret AI recruiting tool that showed bias against women. In *Ethics of Data and Analytics* (pp. 296–299): Auerbach Publications.
Eubanks, V. (2018). *Automating inequality: How high-tech tools profile, police, and punish the poor*. Martin's Press.
European Union. (2021). Proposal for a Regulation of the European Parliament and of the Council Laying Down Harmonised Rules on Artificial Intelligence (Artificial Intelligence Act) and Amending Certain Union Legislative Acts. https://eur-lex.europa.eu/legal-content/EN/TXT/?qid=1623335154975&uri=CELEX%3A52021PC0206
Gebru, T. (2018). How Can We Stop Artificial Intelligence from Marginalizing Communities? TED. https://www.ted.com/talks/timnit_gebru_how_can_we_stop_artificial_intelligence_from_marginalizing_communities
Langley, H. (2020). One of Google's leading AI researchers says she's been fired in retaliation for an email to other employees. Business Insider. https://www.businessinsider.com/timnit-gebru-ethical-ai-fired-google-2020-12
Latour, B. (1990). Technology is society made durable. *The sociological review, 38*(1_suppl), 103–131.
Martin, C. D., Huff, C., Gotterbarn, D., & Miller, K. (1996). Implementing a tenth strand in the CS curriculum. *Communications of the ACM, 39*(12), 75–84.
Moyser, M., & Burlock, A, (2018). Time use: Total work burden, unpaid work, and leisure. Statistics Canada. https://www150.statcan.gc.ca/n1/pub/89-503-x/2015001/article/54931-eng.htm

Perez, C. C. (2019). *Invisible women: Data bias in a world designed for men.* Abrams.

UNESCO. (2021). *Recommendations on the Ethics of Artificial Intelligence.* https://en.unesco.org/artificial-intelligence/ethics

United Nations. (1948). Universal Declaration of Human Rights. https://www.un.org/en/about-us/universal-declaration-of-human-rights

Sadeghi, S. (2018). Ancient Women Mathematicians. Presentation given at INVIDI Technologies Corporation. https://www.birs.ca/workshops/2018/18w2043/files/Sadeghi.pdf

Sharma, K. (2019). How to keep human bias out of AI. TED. https://www.ted.com/talks/kriti_sharma_how_to_keep_human_bias_out_of_ai#t-78285

Tétreault, M. A. (2001). Frontier Politics: Sex, Gender, and the Deconstruction of the Public Sphere. Alternatives: Global, Local, Political.

West, M. K., Kraut, R., & Ei Chew, H. (2019). I'd blush if I could: closing gender divides in digital skills through education. *UNESCO and EQUALS Skills Coalition.*

# Part I
## Work

The first part of this book will examine three facets of work in the digital age—the right to work, care and unpaid labor, and leadership. Women have always worked but have often faced barriers to being remunerated for their labor. In the last two hundred years, we have seen in most countries of the world a progressive opening up of women's rights in relation to work, whether it be accessing certain professions, obtaining decent pay and working conditions, and becoming leaders and decision-makers in the public sphere.

As we will see, care work (much of which is underpaid or unpaid) is one of the last bastions of women'swork in many parts of the world, where the responsibility for taking care of young children and the elderly vastly falls on women. The effect of this is the propagation of stereotypes that claim that women are "better" at this type of work than men, leading to men being discouraged to pursue these activities. Research shows that women who engage in other types of careers often still must perform the majority of care tasks in a household or family, which can their ability to develop professionally.

While we are seeing an increasing number of women in the public sphere, in positions of corporate, academic, or political leadership, public perceptions are still different for women in power than for men. We see on the one hand statements that women are "better" than men at leadership, while on the other there is much less tolerance for error in judgment. In fact, women should be leaders simply because they are human beings, and there will be some great leaders and some not-so-great, just like men.

All of this comes against the backdrop of two phenomena. First, we have the reversal of women's rights in relation to work in certain countries around the world, such as Afghanistan. As this book was being written, women in Afghanistan were being banned from public-facing work and from educational institutions, and asked to cover themselves completely when they left their homes. Similarly, the reversal of women's reproductive rights in the United States, Poland, and other countries is certain to have an adverse effect on employment outside of childcare. After all, having children is an enormous, expensive, and very time-consuming commitment, particularly for the mother.

There are several innovations in the age of AI that are transformative for women's rights at work, and others that can be very harmful. In the positive, we certainly see the increase in opportunities for remote and flexible work, which allow for both working parents to be more present and active around their children, especially during their early years. The last few years has normalized the presence of children on work conference calls, a trend that helps to remind us of why we all work so hard after all (hint: it's often to provide for the people we love and create a better future for the next generations).

My biggest concerns for women's rights at work in the age of AI, however, are related to automated decision-making systems, recommendation systems, and generative AI. Automated decision-making systems, such as algorithms that rank resumés automatically to select a shortlist for a job, have been shown many times to discriminate against female candidates, risking to further inequalities in the work force in terms of access to employment and remuneration. Recommendation systems perpetuate content, whether it is on social media, streaming platforms, job sites, or news sites, that appears "more likely" to appeal to the person accessing them. Women disproportionately are shown content around relationships, parenting, and beauty, while men see much more content linked to business, sports, and finance. This content creates realities in which women and men develop expectations about themselves and their gender which opposes any evolution or modernization. Similarly, generative AI is a category of AI that creates text or images based on certain prompts. Study after study shows that the tools currently available perpetrate gender stereotypes, writing texts with less female protagonists or heroes, sexualizing women, or promoting more men in positions of financial and political power.

So, even in the age of AI, battles around gender and work have not changed so much in the past two hundred years. The battlefield, if I may, seems quite different, with challenges related to algorithms, virtual calls, and all kinds of new modes of living. However, the core issues, from allowing women to truly be an equal participant in the public sphere, and fostering men to participate fully in care work, are still the same. In the following three chapters, we will explore these themes in greater detail and dive into the complexities of work in the digital age.

# 2

# Woman Is to Housewife as Man Is to Programmer

*Fresh out of university, a young woman applies for work in several large software companies, hoping to obtain experience and begin to pay off her student loans. She studied computer science and economics and graduated in the top 5% of her class. She also was the captain of the varsity football team, no small feat. However, after several weeks of applications, she has received no call-backs at all. Confused, she re-evaluates her options. Others in her class seem to be getting interviews, but for reasons she can't understand, she is not getting through. What she doesn't know is that her athletic experience is holding her back. The algorithms used by these companies have filtered out phrases in her resumé that correspond to female activities, notably that she was the captain for the women's football team. She is being rejected before a human being even sees her application. She doesn't realise this, however, and starts to believe that she is the wrong fit for this type of organisation, eventually going for a lower-paying job in a different sector instead.*

"Imagine a young girl doing research for a school project," says Kriti Sharma, an artificial intelligence researcher with Human Rights Watch. "She Googles the word CEO, and she gets images of a man. Then she does a search for secretary and, as you might expect, she gets images of mostly women. And now she wants to listen to music and order food, and she barks orders to an obedient female voice assistant. What impact does this have on the way she sees herself, and on the opportunities that she believes she deserves to pursue?"

Of course, women have always been a critical part of the workforce. Before the industrial revolution, they actively participated in the workforce as agricultural workers, textile manufacturers, and craftswomen. By the 1800s, they were commonly employed in factories, as both skilled and unskilled workers.

However, women have faced many constraints related to work over the past two centuries. Women were barred from the legal profession, for example, until the late nineteenth century, and in some cases well into the twentieth century. Women also continue to fight for decent conditions at work—for equal pay and benefits, on the one hand, and for support for their unpaid care work, on the other. There is also the ongoing issue of discrimination and stereotyping at work, which still bars many women from progressing in their careers, and can make them vulnerable to harassment.

In this regard, artificial intelligence has not exactly been friendly to women. When, in 2018, it came out that Amazon had been using resumé-sorting algorithms that discriminated against them, women started thinking about what clues, aside from their name, there might be in their resumés about their gender. From George Sand to Joan of Arc, women have had to pretend to be men throughout history to access certain jobs. Today, we don't know which companies use which human resources algorithms, and if they have been properly vetted against discrimination. There are very likely many more companies, like Amazon, that downgrade a resumé for being too womanly. Resumés and professional identities are critical to achieving socioeconomic equality. Do we still, like Sand, need to scrub them of our gender?

## 2.1 The Right to Work

The reason for these biased algorithms, we are told, is that the data overrepresents men in certain (better paying) technical fields. As more women enter these fields, for example as engineers, the algorithms will learn to recognize that women can be good coders too. However, this won't happen if female candidates are rejected before they even pass an interview. And we don't know how many times this might have happened since AI first was used for resumé screening—it is now used widely around the world for this purpose, and not only in tech companies. Interestingly, Amazon only has 26% of technical positions occupied by women, the lowest percentage of all the American big tech companies. This certainly feels like a step backwards for women's rights at work.

In the early nineteenth century, women were still excluded from all aspects of the political and legal system in most countries. Women could not vote, run for office, or be lawyers or judges. It was not until the mid-nineteenth century that some universities accepted women into their ranks so that they could train to become professionals. In 1862, Mount Allison University became the first Canadian university to allow women to enter. In the United States, it was Oberlin College in Ohio, as early as 1837. The first woman to practice law in Canada, and in the Commonwealth, was Clara Brett Martin in 1897.

Martin graduated from the University of Toronto with a degree in mathematics at the age of sixteen and applied to the Law Society of Upper Canada for permission to study law. The Law Society argued that, according to its bylaws, only persons were allowed to practice law and that only men could be considered persons. This circular argument is repeated again in biased algorithms, where women are not considered potential programmers because they have not been included in the profession in the past.

Many other laws discriminated against women in the workplace until very recently. For example, until 1955 in Canada and 1964 in the United States, women working in the public service faced a marriage bar, meaning that if they married, they had to immediately resign. Maternity leave—a key element of women's participation in the workplace—appeared in the early twentieth century in Mexico and in the 1970s in Canada. Women in the United States are still waiting. In both Canada and the United States, predominantly female sectors were excluded from the first laws setting a minimum wage. In the United States, it was not until 1974 that service, domestic, and retail jobs, held overwhelmingly by women and people of color, were also included (Table 2.1).

The right to equal and fair employment for women was a central theme of the first (late nineteenth and early twentieth centuries) and second (1960–1980) waves of the feminist revolution. This was primarily related to the legal recognition of women as persons. This included, among other things, the right to vote, to stand for election, to own property, and to divorce. Once this was achieved, women took up the challenge of legislating against discriminatory practices that made it almost impossible to succeed in the workplace.

From an international perspective, it was not until 1948 that women's right to work was recognized through the Universal Declaration of Human Rights. Article 23 of the Declaration addresses many of these barriers. It states that everyone, irrespective of sex, has the right to freedom of choice of employment, fair remuneration, decent conditions, and social protection.[4] The drafting committee for this document was chaired by Eleanor Roosevelt, the First Lady of the United States from 1933 to 1945. She had a prominent writing and political career and was later appointed as the United States representative to the United

Table 2.1 Historical obstacles to the participation of women in the labor market

| Description of law | Canada | United States | Mexico |
|---|---|---|---|
| Women not required to resign once they are married[1] | 1955 | 1964 | N/A |
| Women are allowed to practice law | 1941[2] | N/A | N/A |
| Minimum salary for women | 1918–1960[3] | 1938–1974 | 1917 |
| Maximum work week for women | 1960s | 1938–1974 | 1917 |
| Pension for women outside of teaching | 1975 | 1935–1950 | 1943 |
| Maternity leave paid by state | 1971–15 weeks (Now 15 to 61 weeks) | N/A–0 | 1917–1 month (Now 12 weeks) |

Nations General Assembly in 1946. Several other women also shaped the Universal Declaration, including Hansa Mehta, India's delegate to the UN Commission on Human Rights. She was notably responsible for changing the first sentence of Article 1 from "All men are born free and equal" to "All human beings are born free and equal."

This founding text of the new post-World War II era established that "race, sex, language or religion" cannot be used as a basis for discrimination. Despite this, the United Nations took other measures to fight discriminatory practices against women. In 1979, UN member states therefore adopted the Convention on the Elimination of All Forms of Discrimination against Women (CEDAW). Article 11 of the convention recognizes that discrimination against women is a violation of human rights.[5] The right to be employed outside the home is affirmed, as is the right to equal pay, health and safety.

One hundred and eighty-nine countries have ratified CEDAW. This means that they have a legal obligation to respect it under international law. Only the United States and Palau have signed but not ratified it, which means that they participated in its drafting but are not legally bound by it. Iran, Somalia, Sudan, and Tonga have neither signed nor ratified the treaty.

By 1979, most of the world's countries had thus recognized that discrimination against women in the workplace and an unequal division of labor in the home was a violation of human rights. For even if women had the right to work outside the home, and lead organizations and governments, they had neither the time nor the leisure to do so, so great was the pressure to perform domestic tasks. Allowing women to participate fully in the world of work therefore also meant reducing their domestic duties.

As with many international treaties, however, a clear and binding implementation strategy was missing. In 2000, therefore, a protocol was put in place to allow individuals and groups to lodge complaints about violations of CEDAW. The Committee on the Elimination of Discrimination against Women was established to monitor the implementation of the convention. This group of experts receives reports of alleged violations, conducts investigations, and makes recommendations. There is also the Commission on the Status of Women (CSW), which convenes yearly in New York as the main organ to promote gender equality and women's rights. Every year, it focuses on a different theme—in 2023 it was, appropriately, "*innovation and technological change, and education in the digital age for achieving gender equality and the empowerment of all women and girls.*"

Following the adoption of CEDAW, there have been other international initiatives on women's rights, including three other women's conferences (the first was held in Mexico City in 1975 to prepare the ground for the development of CEDAW)—in Copenhagen in 1980, Nairobi in 1985, and finally, Beijing in 1995.

## 2.2 Stifled by Stereotypes

Today, algorithm or not, many women face the same stereotypes in the workplace as did their mothers and grandmothers. Even now, with a much stronger legal and international support structure than ever before, the women I spoke to for this book have worked in professional environments that have often held back their careers. Many of them have left, frustrated, and tired of banging their heads against the glass ceiling.

*Ana*, a former employee government employee, is one of them. For several years, she worked long hours in a consulate in the United States. She managed cultural affairs, regularly staying at work until 11 pm to prepare for the next day's events. After several years at the consulate, her supervisor was offered the general management of a government agency in her country. He suggested to her that she leave her job, and the United States, to follow him and become director.

"I came back to Mexico City firmly believing that I would be the director of my department. I was young, but I had almost eight years of experience in the field. When I got there, my supervisor delayed my promotion saying, things have changed a bit. I don't want confrontation. I'm sure people will leave on their own. I was a bit nervous and intimidated, so I agreed."

He told Ana that the person she was to replace was much more competent than she was, and that she would have to wait. She found herself in an assistant role, doing tasks far below her competence. In the meantime, several of her male colleagues were promoted while she waited for a turn that never came. She left the organization and now works as an independent communications consultant.

Ana's story raises two interrelated issues. Firstly, many women still run the risk of becoming the "office housewife." They do administrative tasks, make coffee, while they are hired for another role. This serves to reinforce stereotypes about them that prevent them from progressing professionally. Secondly, frustrated, many women end up quitting and becoming self-employed.

In a 2015 article entitled: *Madam CEO, get me a coffee*, Adam Grant and Sheryl Sandberg describe the similar experience of a woman in New York. They note that: "This is the sad reality in workplaces around the world: Women help more but benefit less from it."

The first professional jobs assigned to women confined them to roles where they assisted men. They were mainly found in clerical positions (typing, sending mail, organizing calendars) where they served the needs of others. Although many other areas of work are now open to them, women are still often confined to these types of tasks, reminiscent of those they performed in the past and at home. A surprising analysis of the 2010 United States Census estimates that jobs for women have not changed much since the 1950s. In 1950, the most common jobs for women were: secretary, bank teller, salesclerk, domestic worker, and teacher. In 2010, they were: secretary, cashier, teacher, nurse, and nurse's aide.

My grandmother graduated from Smith College, a prestigious liberal arts college in Massachusetts, and spoke three languages. An American from Chicago with a passion for politics and history, she worked for the OSS (Office of Strategic Services), the equivalent of the CIA during World War II, before marrying a Canadian businessman and moving to Montreal. As an intelligence analyst, specializing in Latin America, she was bound by state secrecy and never broke her oath. She therefore spoke very little about her experience in the secret service. The only anecdote she allowed herself, which still outraged her decades later, was being asked by her superior to pick up his laundry from the cleaner. From what I remember, she refused to obey this order, which probably did not help her relationship with her superior. Later, she asked to be sent on a mission to Peru, but her request was rejected. According to family legend, she was told that not only was she a woman, but that she was too tall (over six feet) and would attract too much attention as a spy. A few years later, as married women still had to do in the late 1940s due to the marriage ban, she left the civil service to start a family. For the rest of her life, however, she remembered how sexist and disappointing that environment had been. Settling in Quebec, she nurtured a life-long passion for learning, but never achieved her career aspirations.

Three decades after my grandmother's experience, in 1977, Professor Rosabeth Moss Kanter of the Harvard Business School published the book entitled *Men and Women of the Corporation*. In it, she describes how, at that time, women were trapped in a spiral of personal favors to their superiors: "…the only way they can get recognition and perhaps advancement is to develop a personal supportive relationship with their boss, only to find themselves further tied to that boss and rewarded for things that are not necessarily useful elsewhere in the organisation."

This form of subserviency to a supervisor still exists. Millions of women—educated, passionate, full of enthusiasm for their profession—have been pushed into tasks far below their skill level and far from their aspirations. In many ways, this tendency is accelerated by AI, notably by natural language processing models, which perpetuate the type of employment we associate women with. These models are usually used to inform online text, such as that produced by translation algorithms, chatbots, social media posts, and even news articles.

In a striking 2016 paper, it was found that the associations embedded in these models contained strong gender stereotypes. For example, female pronouns were associated with terms such as housewife, nurse, and receptionist, while male pronouns were associated with terms such as philosopher, captain, and architect. In practical terms, this means that in technologies that use such models, the text or computer-generated conversation makes very gendered assumptions. There are, however, more ways in which existing stereotypes and AI conspire to perpetuate norms that women like my grandmother already faced, without allowing our society to evolve.

## 2.3 COVID-19 and the Second Shift

The COVID-19 pandemic was extremely difficult on working mothers. Already facing an enormous workload with their careers and their second shift at home, they now had the added double burden of a looming economic crisis and no school or childcare.

When, in early March 2020, a week of school closures turned into two, then six, parents adapted. Women in particular, who already had the most responsibilities at home, were overwhelmed by the pressure. Thousands upon thousands of articles were written on the subject, but governments worked with a limited set of tools—open schools; or close them. There didn't seem to be many possible alternatives, of allowing families to combine forces, for example, of allowing nannies, or even of partially funding babysitters.

Consequently, many women, unable to keep up with their jobs while caring for their children full-time, started slipping out of the workforce. In June 2020, the New York Times reported that working women in the United States were starting to quit their jobs. In July, the news blog *fivethirtyeight* chimed in, arguing that women were already completely overloaded with domestic duties, and that this increase by 40 hours a week (the time children spend at school) was causing them to either quit or face burnout.

In its report on women and COVID-19 in Canada, the YWCA noted that women had accounted for 63% of job losses in May 2020, at the height of the first wave of the pandemic, and that the jobs returned twice as fast for men (2.4% increase) as they did for women (1.1% increase). In the United States, in one month, women lost all of the gains that they had made in employment in the past decade.

How is that possible? I interviewed working mothers to find out. Emma, a new mother working in knowledge management for a law firm in Toronto, spoke to me in April 2020: *"work-life balance with a young child and being a full-time professional was already challenging before Covid - and I don't know if balance is even achievable - and that idea has been totally abandoned now that we are in the thick of this crisis - working, parenting, and trying to manage everything within our home. Before I was a bit sceptical and even rejected the idea of balance, but now I don't even know if it is possible."*

While she answered my questions thoughtfully, she, like every other mother of young children that I knew, was visibly exhausted. Having recently returned from maternity leave, she was working in quarantine with an eighteen-month-old boy who was still breastfeeding.

*"The schedule is completely unsustainable; I don't know how we have lasted this long."* At that time, she was 46 days (by her count) into a routine that would last many more months. With her husband, she worked in shifts, non-stop, from 6:00 AM to midnight or 1:00 AM in the morning. Very often, work extended into the weekend. *"Taking care of a toddler isn't a break from work. It's in many ways more work than what you actually get paid for. It feels like we are doing the impossible. Sometimes I feel like my head is just above water."*

These feelings were echoed by numerous working mothers, particularly those with young children. Overwhelmingly, they said that they did more domestic work than their spouses, were more exhausted, and even when they were the primary breadwinner, weren't able to find time for their careers. Many women spoke about mental health issues and felt that they would not be able to last more than a few more months into this routine.

In the first wave of the pandemic, while mothers of young children held their breath, decision-makers went back and forth between impossible options—open the schools and risk a second wave of the virus, or close them and risk parental breakdown. When governments asked parents to keep their children at home to help stem the pandemic, there was only limited support, or alternative and innovative help that might make this proposition a little easier. And because women are already the primary childcare provider in the majority of families, this additional burden went on them.

"Working from home is not what is happening here," said Emma. "Parents are in their homes, and have been asked to bring their place of work into their homes, but also care for their children and be their teachers. We have been asked to do what is an impossible task. It is impossible to ask people to do all of these things all at once."

As long as domestic duties continue to be assigned primarily to women, there will be the motherhood penalty. This statistic measures the percentage of income lost, compared to a female peer without children, for each children a woman has. This compounds the gender pay gap, which measures the difference in pay between men and women for equivalent duties.

In the United States, the New York Times reported that women get a 4% pay cut for each children that they have, compared to a 6% pay increase for men. This is in part due to parental leave, which, while very important, is still taken mostly by women. In the book Fair Play, Eve Rodsky writes that the imbalance in domestic duties actually increases at home when women have children. So, even in cases where men were taking on more duties, once a baby is born, women end up doing even more.

In work environments that don't take into account families at all, it is unsurprising that men hesitate to take on more of the house and care work. Having taken two maternity leaves in recent years, I can say without hesitation that both times, I fell behind at work and struggled to get back on track when I returned. The months or years off that women take after having children can have a dampening effect on their careers, which is often accentuated by continuing childcare responsibilities afterward.

One of the best pieces of advice I ever received was from Monique Jerome-Forget, former Quebec Minister of Finance, who told me that: "When you return from maternity leave, you should always ask for a promotion. After all, you performed a great service to the state. Where would we be if women stopped having children?"

The reality, however, is that as we continue to penalize women for having children. Instead of thanking them, we contribute to the discrimination of women at work. As Emma notes, it is impossible to ask women to do all of these things all at once. Ultimately, the burden of childcare still falls on women, and as long as our society continues to take this for granted, women will be asked to compensate for all crises, big and small. Child is sick? Mom takes off work. Snow day? Mom takes off work. Global pandemic? Don't worry, mom is here.

How could we possibly be surprised that women are struggling to break the glass ceiling in a society that asks so much or them?

## 2.4 The Family or Career Paradox

A few years ago, after moving to a small town on the North Shore of Montreal, I visited the local library with my family. I came out with Sleeping Giants, the first volume of a science fiction trilogy by Sylvain Neuvel. Back home, I plunged into my book. A young girl, Rose Franklin, falls into a giant metal hand buried in the ground. She grows up to become a physicist. In search of the meaning of her discovery, she will lead a team to find the other body parts of giant robots, scattered around the world.

Rose is a somewhat stern, intellectual, and introverted woman. She grew up in her books and her imaginary world. She eventually becomes a model of feminism with a fulfilling professional life, while remaining true to herself. The character of Rose Franklin is likely inspired by Rosalind Franklin, the English chemist who played a central role in understanding the structure of DNA. In this sense, Sleeping Giants is a tribute to the brilliant women of history.

Rose also resembles the character of Eleanor Arroway from Carl Sagan's novel Contact. They unfortunately also have their social isolation in common. Disappointed by humans, they concentrate on their work and reject proximity to their fellow human beings.

Sylvain Neuvel and Carl Sagan, probably unconsciously, use a very common stereotype in our society: that of the heroic, brilliant, and lonely woman. Women are rarely seen as both competent and warm. The extremely intelligent women in these novels do not have children. They set the tone: mothers cannot succeed professionally in the same way that women without children or men can. Male characters, on the other hand, are often fathers.

A study by Catalyst, an organization that works for the position of women in the professional world, speaks of a major dilemma for women in management positions. It shows that gender stereotypes create a lose-lose situation for women. According to the study, men and women tend to associate leadership with men. When women show leadership, they are seen as less friendly, even cold. On the other hand, when they are caring and empathetic, they are more liked by their colleagues, superiors, and subordinates, but are also seen as less competent leaders.

It is easy to see why the characters of Rose Franklin and Eleanor Arroway have no children and no friends. They are the competent leaders in our narrow collective imagination. Of course, these perceptions are wrong. Many extraordinary and innovative women leaders are also caring and devoted mothers, wives, and friends. However, they may feel trapped in this double bind where they are seen as either not competent enough or not warm enough, with no middle ground.

These types of stereotypes, however, are not just propagated by literature. Recommendation algorithms are immensely guilty, particularly those that share online text or video content. For example, I feel particularly stuck by my Netflix recommendations, which always give me the same type of content: romantic comedies and shows about professional mothers. I noticed the narrowness of the diversity of what was offered to me, when I looked at the recommendations on my husband's profile. Why were science fiction and fantasy films never offered to me? This silo in which I am locked forces me to actively search for content if I want to go beyond what is being proposed to me as my options.

An AI-powered recommendation system locks women into these types of stereotypes, making it difficult for her to browse through other options, as she doesn't know that they exist. This constant association of stereotypes in relation to the type of life she should be living will influence the work that she believes is accessible to her, and limit what she perceives as her choices.

## 2.5 Multilateral Efforts against Discrimination

The 189 countries that signed CEDAW did not just commit to allowing women to work. In Article 5[6] of the convention, they committed to revolutionizing the place of women in society. Ms. Anne Lévesque, professor at the University of Ottawa Faculty of Law, calls the article "very radical.

It says that it's not the woman who is the problem, it's the whole system that has to change and we have to destroy the patriarchy."

"Article 5 doesn't tell women that they can leave a little earlier to pick up their children from daycare, it asks countries to completely rethink the work week, to rethink the division of labour, to rethink everything."

This convention requires a commitment to "modify the social and cultural patterns of conduct of men and women, with a view to achieving the elimination of prejudices and customary and all other practices which are based on the idea of the inferiority or the superiority of either of the sexes or on stereotyped roles for men and women," as well as to ensure "the recognition of the common responsibility of men and women in the upbringing and development of their children."

However, even with this radical international commitment to the status of women, international mechanisms depend on the willingness of individual countries to actually end discrimination against women. To explain in more detail how this might work, I want to take the example of an evaluation of Canada by the CEDAW committee. The latest recommendation for Canada from the United Nations CEDAW Committee in 2016 deals, among other things, with the implementation of Article 5. The Committee makes the following remark about employment:

"There is still a significant gap between men and women in terms of employment rates, as well as a persistent occupational segregation and wage gap, and that women belonging to minority groups, including migrant women, indigenous women, women with disabilities and young women, are still predominantly employed in part-time jobs and in traditional low-paid jobs." It goes on to say "Please provide information on the measures taken to address these problems in order to ensure equality between women and men in the labour market." If Canada's report did not address this issue at all, it seems to be a significant omission. However, the Committee's recommendations remain vague. The onus is on member countries to come up with quantifiable targets and measures to monitor their achievements.

The Feminist Alliance for International Action (FAFIA) is an organization that produces its own reports on the implementation of CEDAW in Canada. Its findings are clear. The last decade analyzed (2006–2016) was an intolerable setback for Canadian women. Many social programs and monitoring mechanisms for women's rights disappeared. As a result, Canada slipped in the United Nations' Gender Inequality Index (GII) global ranking. It slipped from 16 to 25th place out of 162 countries between 2008 and 2014. In 2021, it managed to climb back to 17th place. Once again, this situation shows the fragility of women's gains.

Mexico is 75th in the Gender Inequality Index in 2021. It was 68th in 2010, when the GII was first published. Beyond the situation of women in the workplace, the epidemic of violence against women in the country is one of the main causes of this lower score.

Since the beginning of the GII, the United States has hovered between 41st and 53rd place. It has not ratified CEDAW, although the Senate has held hearings on it five times—in 1988, 1990, 1994, 2000, and 2010. As a result, it does not report to the Committee and does not receive any recommendations to improve the situation of women in its territory.

Many women's rights organizations—such as the Prosperity Project in Canada—have recommended monitoring statistics such as the GII, as well as statistics collected at the national level, in order to measure the progress of women's rights. Although treaties leave room for interpretation, it is certainly not difficult to track such things as women's labor force participation, the pay gap, and the maternity penalty. These statistics, collected by different governmental and non-governmental bodies, must be part of a comprehensive and measurable strategy for the elimination of discrimination against women.

## 2.6 Reprogramming Gender Rights at Work

Gender equality at work has operated on three main fronts: the right to work and to choose one's profession; the right to better working conditions—including social security, childcare, and maternity leave; and the right to be free of stereotyping while applying or working.

These priorities are all present, in one way or another, in international commitments on women, including CEDAW. On the first front, it would be critical to review all algorithms involved in resumé sorting, particularly in software companies, to ensure that there are no more exclusion criteria based on gender. Additionally, an audit of the companies having used these algorithms for several years and the possible impacts on gender balance in the companies could be conducted, with possibly accompanying human rights sanctions.

The second front can be addressed through more inclusive practices in software companies, in order to attract and retain female talent, but also in the recognition of the value of work largely held by women, increasing pay and benefits in these areas. On the third front, it is important to retrain algorithms that have a tendency to devalue women's contributions, or "assign" them to only certain types of work.

Finally, norm-setting conventions such as CEDAW may need to be updated to include considerations from the age of AI. This will influence the resources that signatory countries place on AI gender biases at work. Without this, current initiatives to decrease gender bias in AI may lack international legitimacy.

With this chapter, I really hoped to drive home the idea that gender equality at work has a long history and context. Understanding biases and stereotypes in AI and other digital technologies is important, but so is acknowledging that the issues that are arising have come up again and again. At the core of all this is a fundamental tension in gender roles, with certain implementations of AI, if we let them, at risk of pulling us backward.

## 2.7 Suggested Discussion Questions

1. What could we change in an international convention such as CEDAW to make it more relevant to the age of AI? Would adding an article be enough, or should we rewrite the whole convention? Do we need any modifications at all?
2. What policies still need to be put in place in your country to support women at work? What are the barriers to enacting them, how might they be surmountable?
3. How might stereotypes about gender and work influenced you or people you know? Have you observed or experienced discrimination? How was this dealt with?

# Notes

1. This provision was implemented in Section 213 of the US Economy Act of 1932, which provides for the dismissal of one person from each married couple working for the government. Since women were not considered the breadwinners of the family, they were the ones who, in the vast majority of cases, lost their jobs. https://www.history.com/news/great-depression-married-women-employment.
2. This measure was lifted later in Quebec than in other provinces.
3. This measure was lifted at the provincial level, first in Alberta and Manitoba, and then in Prince Edward Island, which covered both men and women. https://www.canada.ca/en/employment-social-development/services/labour-standards/reports/federal-minimum-wage.html.
4. Universal Declaration of Human Rights, United Nations, 1948. Article 23: *(1) Everyone has the right to work, to free choice of employment, to just and favorable conditions of work and to protection against unemployment. (2) Everyone, without any discrimination, has the right to equal pay for equal work. (3) Everyone who works has the right to just and favourable remuneration ensuring for himself and his family an existence worthy of human dignity and supplemented, where necessary, by other means of social protection. (4) Everyone has the right to form and join trade unions for the protection of his interests.*
5. Convention on the Elimination of All Forms of Discrimination against Women (CEDAW), United Nations, 1979. Article 11: *States Parties shall take all appropriate measures to eliminate discrimination against women in the field of employment in order to ensure, on a basis of equality of men and women, the same rights, notably (a) The right to work as an inalienable right of all human beings; (b) The right to the same employment opportunities, including the application of the same criteria for selection in matters of employment; (c) The right to free choice of profession and employment, the right to promotion, job security and all benefits and conditions of service, and the right to receive vocational training and retraining, including apprenticeship, advanced vocational training and continuing education; (d) The right to equal remuneration, including benefits, and to equal treatment in respect of work of equal value, as well as to equal treatment in the assessment of the quality of work; (e) The right to social security, particularly in the event of retirement, unemployment, sickness, invalidity, old age and other incapacity to work, as well as the right to paid leave; (f) The right to protection of health*

*and to safe working conditions, including the safeguarding of the function of reproduction.*

6. Article 5: Les États parties prennent toutes les mesures appropriées: (a) Modifier les schémas et modèles de comportement socioculturel de l'homme et de la femme en vue de parvenir à l'élimination des préjugés et des pratiques coutumières, ou de tout autre type, qui sont fondés sur l'idée de l'infériorité ou de la supériorité de l'un ou l'autre sexe ou d'un rôle stéréotypé des hommes et des femmes; (b) Faire en sorte que l'éducation familiale comporte une bonne compréhension de la maternité en tant que fonction sociale et la reconnaissance de la responsabilité commune de l'homme et de la femme dans l'éducation et le développement de leurs enfants, étant entendu que l'intérêt des enfants est la considération primordiale dans tous les cas.

# References

Adami, R. (2018). *Women and the universal declaration of human rights* (Vol. 32). Routledge.

Barone, E. (2020). Women were making historic strides in the workforce. Then the pandemic hit. *Time Magazine.* Accessed at: https://time.com/5851352/women-labor-economy-coronavirus/

Bolukbasi, T., Chang, K. W., Zou, J. Y., Saligrama, V., & Kalai, A. T. (2016). Man is to computer programmer as woman is to homemaker? Debiasing word embeddings. *Advances in neural information processing systems, 29,* 4349–4357.

Catalyst. (2007). *The double-bind dilemma for women in leadership: Damned if you do; Doomed if you don't.* https://www.catalyst.org/research/the-double-bind-dilemma-for-women-in-leadership-damned-if-you-do-doomed-if-you-dont/

Cohen, W., & Myers, R. (1950). 1950 Social Security Amendments. *Social Security.* https://www.ssa.gov/history/1950amend.html

DeWitt, L. (2010). The decision to exclude agricultural and domestic workers from the 1935 Social Security Act. *Social Security Office of Retirement and Disability Policy.* https://www.ssa.gov/policy/docs/ssb/v70n4/v70n4p49.html

Government of Mexico. (1917). *Diario Oficial.* https://www.scjn.gob.mx/sites/default/files/pagina-micrositios/documentos/2016-12/00130029.pdf

Government of Mexico. (2016). *Incapacidad por la maternidad.* http://www.imss.gob.mx/sites/all/statics/maternidad/pdf/e-book-incapacidad-maternidad.pdf

Government of Canada. (1985). *Canada Labour Code.* https://laws-lois.justice.gc.ca/eng/acts/L-2/page-35.html#h-342149

Government of Canada. (n.d.). *Employment Insurance maternity and parental benefits.* https://www.canada.ca/en/employment-social-development/programs/ei/ei-list/reports/maternity-parental.html#h2.1-h3.1

Grant, A., & Sandberg, S. (2015). Madame C.E.O., get me a coffee. *New York Times.* https://www.nytimes.com/2015/02/08/opinion/sunday/sheryl-sandberg-and-adam-grant-on-women-doing-office-housework.html

Kanter, R. M. (1977). *Men, Women and the Corporation.*

Kurtz, A. (2013). Why secretary is still the top job for women. *CNN.* https://money.cnn.com/2013/01/31/news/economy/secretary-women-jobs/index.html

Marshall, K. (2003). Benefitting from extended parental leave. *Statistics Canada.* https://www150.statcan.gc.ca/n1/pub/75-001-x/00303/6490-eng.html

Morsink, J. (1991). Women's rights in the Universal Declaration. *Human Rights Quarterly, 13,* 229.

Miller, C. C. (2014). The motherhood penalty vs. the fatherhood bonus. *The New York Times.*

Neuvel, S. (2016). *Sleeping giants.* Del Rey Books.

Public Service Alliance of Canada. (2013). *Women's history.* https://www.psac-ncr.com/wp-content/endurance-page-cache/canadian-womens-history/_index.html

Sharma, K. (2019). How to keep human bias out of AI. *TED.* https://www.ted.com/talks/kriti_sharma_how_to_keep_human_bias_out_of_ai#t-78285

Thomas, H. (2021). Women who dressed as men and made history. *Library of Congress Blog.* https://blogs.loc.gov/headlinesandheroes/2021/03/women-who-dressed-as-men-and-made-history/

United States Government. (1975). Social Security Abroad. *Social Security Administration.* https://www.ssa.gov/policy/docs/ssb/v38n8/v38n8p34.pdf

UNDP Human Development Report Office. (2022). *Gender Inequality Index.* http://hdr.undp.org/en/content/gender-inequality-index-gii

UN Women. (2023). *Commission on the Status of Women 67.* https://www.unwomen.org/en/csw/csw67-2023

# 3
# Rebalancing the Distribution of Unpaid Duties

*For the past 10 years, Jade has been working as a data scientist for a big software company. In her mid-30s, she is now managing a small team and just got married to Alex, who works in cybersecurity for a start-up. The pair decide to grow their family, and Jade becomes pregnant. They both live in Seattle, where Jade does not have state-sponsored maternity leave, but will get 12 weeks of full pay, followed by 12 weeks of unpaid leave. Alex doesn't have any parental leave through his work and isn't sure if he can take care of an infant alone. He therefore takes two weeks of vacation when she gives birth and returns to work full-time. After her 12 weeks of paid leave, Jade doesn't feel comfortable taking unpaid leave for a variety of reasons, she is worried that she will lose her place at work, she is worried about the lack of salary, given high housing costs, and she is lonely. Jade finds a daycare for her baby, but it is a 45-minute drive away, in traffic. Back at work, she pumps milk several times a day, while managing her team. Alex starts taking on more work hours to help with the cost of daycare. After 3 months, of pumping at work, a very long commute, and sleepless nights with an infant that she barely sees during the day, Jade is exhausted and decides to take an extended leave.*

© The Author(s), under exclusive license to Springer Nature Switzerland AG 2023
E. Fournier-Tombs, *Gender Reboot*,
https://doi.org/10.1007/978-3-031-41390-2_3

I don't remember thinking much about gender roles until I had children. For many of the women of my generation growing up in Quebec, feminism seemed to be the struggle of our mothers and grandmothers—gender equality was already achieved, wasn't it? It wasn't until I was in my thirties that I realized that I absolutely had to join the third wave of feminism. Today, however, I am often asked by young female students how to juggle children and work. By their early twenties, many of them are worried—they've been told that high-powered academic or corporate careers don't mesh well with pregnancy and children, and they are wondering if having it all is even possible. In a way, this impresses me, because they seem better prepared for the working world than I was at their age (I don't think I was asking any of those questions). Yet, it also saddens me because I worry that they are already beginning to lower their career ambitions for fear of not making it.

Recently, I gave a talk on women and work and talked about the importance of sharing caregiving tasks equally between women and men. Whether in the professional field or at home, there needs to be as much emphasis on men and care work as there is on women in technology. However, many participants, all women, disagreed. "My husband is a good father," one woman told me with a laugh, "but he would be a terrible mother!" A young woman in her early 20s added, "women have qualities that men don't, and they will always have more work to do at home." My hear sunk. Isn't saying that men are unable to perform care duties just as sexist as saying that women can't be good programmers, or leaders? Most participants believed that women were naturally more empathetic and organized. The solution to women's domestic overload was therefore to help them become more organized, not to promote equal involvement of men. I, for one, believe that this train of thought is harmful to both genders. And I think we too often allow ourselves, as women, to tell the men in our lives that they are not as "good" as we are at caring for others. Which means, of course, that we need to do it all ourselves. Only there are just so many hours in a day.

Unfortunately, AI seems to be partial to very strict gender roles too. A study conducted with Céline Castets-Renard in 2021 (verified in 2023 to be still the case) shows blatant examples from Google and Microsoft Bing translation algorithms. When given phrases to translate in a language that

Table 3.1 Translating non-gendered pronouns using AI

| Original Pinyin | Google Translate – English | Microsoft Bing Translator – English |
|---|---|---|
| Ta zhengzai touzi | He is investing | He's investing |
| Ta shi zhexue jia | He is a philosopher | He is a philosopher |
| Ta shi chengxu yuan | He is a programmer | He is a program dollar (translation error) |
| Ta shi mishu | She is a secretary | He is the secretary |
| Ta shi duizhang | He is a captain | Unable to translate |
| Ta zai xi yifu | She is doing laundry | He's washing clothes |
| Ta zai zhaogu haizi | She is taking care of the child | He's taking care of the kids |
| Ta zai gongzuo | He is working | He's at work |
| Ta zai kaiche | He is driving | He's driving |
| Ta zai tiaowu | She is dancing | He's dancing |

does not use gender pronouns (in this example, Pinyin), Google inserts pronouns along traditional gender norms, while Bing ignores the female pronoun altogether (Table 3.1).

In a 2022 CBC Ideas episode entitled: "A Harem of Computers," producer Jill Fellows tried this using Malay instead of Pinyin, with similar results.

Why is this a problem? Well, the AI subfields of Natural Language Processing (NLP)—which means interpreting text, and Natural Language Generation (NLG)—which means creating text, are currently growing at an enormous rate. New language models, such as GPT-4, which is one of the most recent and powerful model today, are used to create all kinds of texts, from news articles, to social media posts, to even Wikipedia entries. GPT-4 is trained on the Internet—a mass of texts found online, which is, of course, full of hateful and misogynistic opinions. The algorithm, not knowing any better, just reproduces what it was fed. And we, often not knowing that a badly trained algorithm wrote them, read texts that refer to women doing childcare and men creating AI, and continue to believe that women are better at childcare and men at programming.

In order to break out of this vicious cycle, we need to be aware of it. But the tricky thing about social norms is that they are taught and reinforced by cues we get from our environment. Most of us spend much of our days on the Internet, reading news and social media posts which may well have been written by algorithms regurgitating whatever garbage they were fed.

This makes things even more difficult for those who want to challenge gender norms around childcare. During the pandemic, I remember calls with exhausted friends and colleagues who said, "My husband's job is less flexible than mine, and he's also less resilient with kids, he gets tired quickly. So, I organize myself to work around his schedule." There was one who was losing her hair. Another one was working all night, sleeping three hours a day (actually, many of us went through this). Without school or the ability to get help, it was imperative to have someone in the household to take care of the children. But women continued to take on the majority of the care workload, not only egged on by their community, but now, increasingly, by AI.

However, the problem with childcare is not just that women are encouraged to drop everything to engage in it. It's also that men that want to care for children—whether as fathers, early childhood educators, or babysitters—face enormous social barriers. Examples of stay-at-home dads exist and are becoming more common. Men who choose to give up part of their career to care for their children are a small but growing minority. These men are often trying something new for themselves, a way of existing and expressing their masculinity that breaks with their family traditions. But they need much more support and encouragement that they are currently receiving.

## 3.1 First Generation Househusbands

Paolo talks to me, through Zoom, of course, a little emotional. He is telling me that when she was three years old, his daughter had an accident, seriously injuring her foot. During the three or four months of physiotherapy, he took care of her, accompanying her to her appointments. His partner Isabel, meanwhile, worked long hours at an airline. When his daughter healed, Paolo and Isabel determined that Paolo would continue to be a stay-at-home dad for their two daughters.

"It allowed me to spend a lot of time with them," he recalls, smiling. "We would go to the park, pick up Isabel from work, do homework, climb trees, ride bikes, go to the pool... and I was at home with them for fifteen years. It wasn't complicated, but we gradually adapted to the situation."

"It's been very rewarding to be there for the girls," he adds. "I don't know what direction my career would have gone in if I had continued, but my daughter's accident changed me, and I realized that you can't pay someone to love your child. That I could be there to fulfill that function was much better." (As a side note, I personally believe that nannies and paid caregivers can provide very high-quality and loving care, but this is not my quote.)

In Mexico and most other countries, the social norm is for men to be the primary breadwinner in the home. Beyond cultural pressures, society itself is still structured according to a traditional model. Indeed, in the government's census form, each family must designate a "head"—the *Jefe de hogar*. The concept of a head of household is controversial because it signals that one person—the one who earns the most money—is also the one who makes financial decisions for the entire family. This is a delicate situation that creates an unequal dynamic between the adults in a family: those who are paid for their work and those who are not.

In most countries, women stay at home much more often than men, although this trend is beginning to change. In the United States, 17 percent of stay-at-home parents in 2016 were fathers, a significant increase from 10% in 1989. Surprisingly, the percentage is lower in Canada where there were only 10% stay-at-home dads in 2015.

It's important to note that financial decision-making and big paychecks don't always go hand in hand. In fact, there are many systems for managing household finances. British researcher Jan Pahl found in 1990 that women tend to control finances in low-income households, while men control finances in higher-income households. In couples who pool their income, women control spending in middle-income households, while men control spending in higher-income households. Wife-controlled pooling was more common when both partners were employed, while husband-control was more common when only men were employed. Yet, the more women contributed to household income, the more likely they were to have a significant say in household spending. According to researchers Carolyn Vogler, Clare Lyonette, and Richard Wiggins, in most households, the larger the expense, the more likely it is that the primary financial supporter will make the final decision—usually a man.

Returning to the government definition of head of household in Mexico—in both Spanish and French, the term commonly used is *jefe de hogar*, in the masculine; for the term to be appropriately used by a woman, the designation should read *jefa de hogar*. From the census form, it can be understood that the power dynamic in the couple is not egalitarian, as the head of the household is presumed to be a man.

In the 2010 and 2020 censuses in Mexico, when a man and a woman lived together as spouses, 85% of the heads of household were men. A woman was more likely to be a head of household if she was separated, divorced, widowed, or single.

There are, however, many circumstances under which a Mexican woman could be head of household while living with her husband. And while these dynamics also exist in the United States and Canada, the systemic and social barriers are far more blatant in Mexico. According to the government, a woman can only be designated as head of household if she earns more than her spouse.

There is such a large status gap between paid work outside the home and unpaid work inside the home. This gap is reinforced by legislation that assumes that pay and gender go hand in hand. Being male means getting paid more and being female means getting paid less. Couples who adopt a different family model therefore have to spend much more energy negotiating new ways of living and sharing tasks with each other. It is much easier for a couple to operate according to predefined norms and changing them requires training, communication, and teamwork.

1973, Ruth Bader Ginsburg, who at the time was a human rights lawyer working for the American Civil Liberties Union (ACLU), defended a man named Stephen Wiesenfeld. Wiesenfeld, who had been a stay-at-home father, was left widowed after his wife, a teacher, died during childbirth in 1972. At the time in the United States, social security benefits were provided to women if their husbands died, but not the other way around, the expectation being that the primary breadwinner in a household would be the man.

Ginsburg argued that the 1935 Social Security Act discriminated against men based on their gender, and that Wiesenfeld, like all men in a similar situation, should be awarded the benefits. Ginsburg won the case and in so doing made an important point about gender discrimination. She argued that cases such as these were important not only in reversing assumptions about women's dependency, but also in showing that systemic gender discrimination affects both men and women.

While many men have benefited from the historical patriarchal system, there are also many who have found it restrictive and even crushing. The restriction of men to the public sphere has meant that they have faced difficulties operating in the household or in paid care work. Many men not only want to be engaged fathers, particularly during the early life of their children, but also might want to become nurses, elderly care workers, early childhood educators, or even midwives. Unless these norms are reversed, women will always have domestic and care work as a primary responsibility, and many men will be left frustrated, wishing they could self-actualize as nurturers.

While discussing this concept with friends, I argued that even the norm of only having teenage girls do babysitting work could be harmful. While many men I spoke to agreed that encouraging teenage boys to care for children would teach them important life skills, the response from women was mixed. Mothers of boys tended to agree with me (as I am also a mother of boys) that it would be great to have more male babysitters, early childhood educators, and nannies. Mothers of girls, on the other hand, told me that men and young boys could not be trusted with girls.

We live in a society which is trying to have it both ways by refusing to educate and encourage our boys to become responsible nurturers, and yet criticizing their reluctance and incapacity to do so once they become fathers. In Canada, women make up over 90% of nurses and elderly care workers, 97% of early childhood educators, and 80% of elementary school teachers. The percentages are quite similar for Mexico and the United States. Typically, this work is lower paid and traditionally associated with women's work. What's important to highlight is that these are all careers that require a fair amount of nurturing. The older the children become, the less hugs they might require from their teacher. This is therefore a signal that, in our society, we consider men to be less competent nurturers than women.

### 3.1.1 Can a Midwife Be a... Man?

Louis Maltais is the first man to practice as a midwife in Quebec. His story is important because midwifery, unlike gynecology (which many men practice), is a practice based on nurture and deep empathy. In his training, he worked hard to understand their pain, their fears, and their excitement.

He was not exactly welcomed with open arms. Maltais faced numerous prejudices entering the profession—from mothers, from fathers, from other midwives, even from other medical professionals such as nurses and doctors. As he embarked on the four-year bachelors' degree in Trois-Rivières, Québec, his professor Lucie Hamelin warned him: "I think that you will have an additional challenge because you are a man."

He countered, describing the progressive discovery of his interest in midwifery: "I did my college degree in natural sciences, not knowing at all where life was taking me. Then I went to circus school, and then I took a course in massage therapy. My teacher spoke about a course that would be for pregnant women, and it was clear to me that I would want to take it. This made me realize that I was really fascinated by the phenomenon of childbirth, this experience that women live - it really brought something out in me."

As he started practicing, patients had mixed feelings about having a male midwife, especially women. "The reasons are not very clear on why we refuse that Louis be present at the birth," says an expectant mother in a documentary film about him. "For the visits, I don't mind, but there is a small doubt that I don't want to have. I don't want to have the smallest doubt." The mother goes on to ask that Louis be replaced by a female midwife during the birth. Her husband, however, has a different take on it: "This is a bit like when women started going into construction. Men who were already there said: "What's going on, she won't make it." It's a little bit like that in this situation. People are wondering what he is doing there. It's normal to see a doctor who delivers a baby, but to see a mid-husband, historically, that's not possible."

Nevertheless, Louis Maltais plunged into his new role with emotion and dedication, at one point walking around with prosthetic breasts and a belly to better understand how it feels (female midwifery students that have not had children also do the same thing). With his fellow students, he practiced pain-relieving positions for labor, such as rolling on a ball or hanging from straps in the ceiling. Today, he is a respected practicing midwife in Quebec. I heard about him from my own midwife, who accompanied me during the birth of my second child, and she gave him glowing reviews.

When I had my second child, I was thrilled by the fact that, in Quebec, I could be tended to outside of a hospital, in what is called a birthing center, only by midwives. Birthing centers are beautiful, usually set up in houses and decorated with plants, Aboriginal medicine wheels, and motivational posters. During labor, each couple has a whole room to themselves, where lights are set low and you can play music, even take a bath. It was for me a completely natural and holistic delivery that was a far cry from the 15-minutes C-section of the first birth.

My experience was mostly a positive one, surrounded by what my husband called a witches' coven—three midwifes that coaxed and encouraged me during those last moments of pushing. This does bear the question, however: if one or two of those witches had been men, would it have been the same experience?

Writing this today, I'd like to say that I would have welcomed a male midwife like Maltais. At this stage of my life, after years of reflection on gender norms and how they negatively affect us all, I probably would have. However, earlier on, when I was just starting to notice discrimination against women in my career, I might have wanted to stake out midwifery as a "woman's turf." After all, some might argue, if we let men into this female space, won't that result in less space for women?

Yet I now feel that just as barriers must be removed for women's entry into areas such as leadership, technology, and finance, barriers for men should also be removed to allow them into nurture and care work. The father who defended Maltais to his wife was right—it is just like construction. Saying that women don't have what it takes to build houses and drive trucks is not so different than saying that men don't have what it takes to empathetically deliver babies.

Today, there are a handful of male midwives in Canada—about 3 in 1,000. All of them take their symbolic role very seriously, speaking to the media or documenting their journey on social media. In the United States, the American College of Nurse-Midwives reports that approximately 0.6% of their midwives are male. Men in Nursing Journal writes of similar barriers for me in the profession, noting: "*Midwifery care focuses on the intimate, intensely personal aspects of pregnancy and childbirth, along with well-woman care. Many believe that a man would be unable to bond with a woman in this way.*"

In Mexico, although official numbers are more difficult to find, there can be a bigger role for men in birthing, particularly in traditional Indigenous cultures. Traditional midwifery practices of the *Me'phaa* or Tlapaneco in the state of Guerrero include a role for men, called *parteros*, or male midwives. There are reports of men in many other Indigenous cultures actively participating in birth, either informally as a family member, or more formally as a *partero*.

### 3.1.2 Fatherhood and Returning to the Roots

In fact, indigenous cultural practices have provided an alternative framework for male nurturers, and not only in the domain of childbirth. Phil, the father of four school-age children, spoke to me about the way in which he combined African, Indigenous, and Polynesian cultures to create a scaffolding for his life as a father and a youth counsellor.

Phil was born in Brooklyn, New York, where he was placed in foster care at a young age. Of African and Cherokee descent, he also has family in Hawaii, which he describes as a Babylon—a mosaic of cultures and languages. To him, diversity means freedom.

Phil spoke about a return to a form of masculinity that was traditionally present in Indigenous and African cultures. "These cultures were rooted in the family," he said. "It was when men became slaves that they had to leave home to work, and they became important in their family lives only as far as they could have financial value to their owners."

This viewpoint was developed by Cheikh Anta Diop, a controversial Senegalese historian who passed away in 1986. In the book Precolonial Black Africa, he writes that African societies were built along matrilineal lines, with leadership roles, property, and clan-belonging inherited by women. While caste systems and slavery were already present in West Africa, colonization imposed specific patriarchal norms on slaves and their descendants in the Americas. These norms were created from a combination of the labor requirements of male slaves, as well as the transmission of a patriarchal culture from Europeans.

Kay Givens McGowan has made similar statements in relation to Southeastern Native American cultures, which include the Cherokee, the Choctaw, the Chickasaw, the Muskogee, and the Seminole. Like Western Africans, these were cultures in which the woman played a leadership role.

"Economically, the Southeastern nations were agricultural people," she writes. "Women farmed and controlled the crops that their work produced, so they were often the traders. This set of responsibilities was confusing to and frequently misunderstood by the early Europeans with whom they traded. In their imposed patriarchy, European men just assumed Native societies were like the male-dominated societies of Europe, in which the economy rested entirely in male hands."

It is important to realize that matriarchal societies were not dominated by women in the same way that patriarchal societies are dominated by men. Typically, these societies were more equalitarian, which means that the onus of being productive and domineering fell on neither gender. Both men and women tended to play an equally important part in the core functioning of the society, and most importantly, both shared in the care work. This means that childcare, elderly care, the care of the sick—these were all spheres in which men were welcomed as well.

Phil has made his career in the community development sector, mainly as a youth advocate and counsellor. He feels that Hawaiian culture, like some African and Native American cultures, may offer better options for a man who wants to be a present father and make a contribution as a social worker. Like most Pacific Islands, Indigenous culture in Hawaii was matriarchal and matrilineal. This does not mean that the culture didn't have its own strict gender norms, but they were different from European ones. The culture also made room for the third gender—the mahu, which applied to people who combined male and female traits.

Speaking about his profession, Phil says: "I got into the work because I was a consumer of various systems myself. Foster care, post-traumatic stress disorder, trauma of abandonment and abuse. I was labelled special-ed because I had fears and would not open up." As a child, he constantly waited to be uprooted and sent to another foster family. Once he was adopted, he remained uncertain and fearful of being "thrown back to the lions."

With this baggage as a motivation, he became a youth advocate for the Mental Health Association (MHA) in New York. He also learned martial arts, which both allowed him to channel his feelings and reach more youth. Needless to say that there are many other men who self-identify as nurturers and have the emotional qualities to be care workers, either professionally or at home. And still today, in Phil and Louis' very different environments (and in many others), they face enormous systemic and cultural barriers. What the two men also have in common is that they have relied on historical knowledge outside of the environment in which they were brought up to teach them how to nurture. Louis in his training in traditional midwifery and Phil in his cultural research both propose that there are different norms for men to draw on and adhere to.

## 3.2 Lean In or Lean Out?

When I was a child, I was only allowed to watch Passe-Partout, which was a French Quebec show that came on for 30 minutes each day. I am fairly certain that every child of my generation grew up as I did, cramped with her siblings on a small couch, enjoying the then-thrilling show that has not aged that well. I also grew up with a nanny who was originally from Iran. She had trained as a nurse and moved to Canada with her husband due to systematic discrimination against her community, the

Bahai. It was with her that I discovered English television, and particularly Sesame Street, which I was allowed to watch on the plastic-covered couches in her living room.

In one episode of Sesame Street, Ernie and Bert repeatedly open and close a door, passing through the doorframe, and singing: "In, in, in; Out, out, out" over and over again. That's the entire skit, and it probably lasts too long (at least for the parents). But I loved it and subsequently spent a lot of time with my siblings singing this song and running in and out of doorways.

I was reminded of this a few years ago when I read Lean In. The book was written in 2013 by Sheryl Sandberg, then Chief Operation Officer of Facebook, and highlighted gender inequalities in the United States while encouraging women to push themselves more in the corporate world. The book was very influential in the United States and was followed by waves of debate, including Marissa Orr's book Lean Out. I felt like I was hearing the jingle all over again. In or out?

I have to admit that I enjoyed Sandberg's book. So much so, in fact, that in 2015 I created a Lean In group in my office, a small tech start-up where 25% of the staff was female. Every woman in the company, from the executive assistant to a vice-president, was a member and actively engaged. We spoke about advocating for our careers, negotiating raises, speaking up at meetings. It was surprising to me at the time to notice that each woman had had similar experiences of discrimination, and it was empowering to try to address them.

After writing her book, Sandberg launched the Lean In Foundation, which encouraged women to push through the sexism in their workplaces and assert themselves. The organization supported groups such as ours with materials, training, and networking opportunities. It even had a star-studded public-service announcement in which Beyoncé appeared, saying: "I'm not bossy. I'm the boss."

On the other hand, those critical of the Lean In movement argued that "playing the man's game" was not necessary in achieving equality at work, and that it was even counterproductive. Melissa Orr, who, like Sandberg, worked at Facebook, said in an interview: "most women don't aspire to occupy the C-or corporate executive level. Research shows that only 18% of women aspire to that. I believe one of the reasons women don't want those jobs is because raising kids and managing a household squeezes their time, so we lose patience and tolerance for the petty power games and office politics. It becomes more important for work to be meaningful." Instead of leaning into the corporate world, which to Sandberg meant being more assertive and dominant, Orr and other critics of Sandberg argued that women shouldn't shy away from qualities that are considered feminine, such as empathy and communication.

In fact, "In" and "Out" can also refer to the public and the private sphere, or the workforce and the household. Both are important parts of women's and men's fulfillment as human beings, and some people do want to lean in, while others, such as the men described in this chapter, need to be supported to lean out.

In short, addressing gender discrimination also means examining the exclusion of men from activities of care and nurture. Many men have attempted to go out on their own, drawing on traditional knowledge, for example, or entering professions in which they are the minority. However, it sometimes feel that in their forays into these areas, they face barriers just as high as women do.

One of Ruth Bader Ginsberg's clerks, Ryan Park, who after his clerkship began his job as a stay-at-home dad, wrote: "men appear to be just as dissatisfied with the stickiness of gender-based norms as women. [...] the emphasis is usually on equalizing burdens—not equalizing the opportunity for men, as well as women, to be there." He argued that most men find their time with their children just as rewarding as women did, and often felt unable to negotiate the time off. When he eventually went back to work, he told potential employers that he would have to leave work at a reasonable time to care for his daughter, just like many mothers do. "This may well have cost me an offer or two." He says, but he felt that trying to achieve work-life balance was worth the effort.

When I hear women saying that men can't be midwives or care workers, or worse, that they can't be trusted with our children, it makes me sad. Sad not only for the men in my life, all of which are naturally empathetic and gentle, but also for the women around me, who will continue to believe that they alone can carry this immense responsibility of care. Without men as babysitters, as nurses, as social workers, as dads, we will continue having the same conversations and talking about the second shift another forty years from now.

Many men in the current generation who are finding alternate ways of operating feel that they are breaking new ground. They often come from more traditional families in which the father primarily acted as the breadwinner or was even absent. In this context, it is challenging for these men to become nurturers without a frame of reference. That is why cultural contexts outside of the North American norm are so important. Whether men are midwives, or paternal social workers, they need institutional and social support to encourage them into these new areas.

Phil's commitment to his family is striking in its contrast to his own father's, who was mostly absent. Speaking to me over the phone while watching his daughter on a scooter, he occasionally calls out to her— "be careful sweetie!". He hopes to eventually leave the mainland United States to take his family to Hawaii. "That is the end game," he says "people are much more family oriented."

## 3.3 Learning to Parent

Diana is a business development manager at an international software company in Mexico City. Her daughter was born just before the COVID-19 pandemic. Just as she was returning to work, the daycare centers closed. Her aunt agreed to take care of the child for part of the day. Then her husband lost his job. He is now developing his business plan to start his own business. This new family context has forced the reorganization of daily life. Negotiating the care that each of them would provide for their daughter did not go smoothly. Like many women, Diana felt she was carrying the entire logistical burden of the child's life.

"My husband often tells me that I have to tell him what to do," she says. "If I don't, he doesn't know that he has to take care of the baby's clothes, take her to the doctor—he has to be told everything."

She adds, "This is my first baby, too. But it's important to me, so I've managed and learned. He has to do the same." Did she learn how to be a mother in the months after her daughter was born, or was she trained throughout her life to know how to organize and manage her home, making her more confident when the time came?

Globally, the vast majority of boys are not raised to be fathers in the same way that girls are raised to be mothers. As we can see from Diana's words, women also do not have the same expectations of their husbands that they have of themselves. Yet, if the problem of male involvement in the home is really a lack of training, the solution seems very simple. It is a gap that some prenatal educators are trying to fill.

Naima Beckles is the director and founder of For Your Birth, a doula agency in New York. Their role is to support couples before, during, and after the birth of a child. She believes that this work should go beyond the traditional childbirth and breastfeeding visits.

"When we are thrown into a situation like this [becoming a parent], we often do what we saw our parents or our communities do. So I see myself as someone who connects our experiences and knowledge to what we are going through now. I teach fathers to compensate for what they don't already know, because many of us [women and men] have not explicitly learned how to parent."

Naima was our doula when our first son was born. She played a very important role in preparing us as parents. During the first meeting, sitting with three other couples on metal folding chairs, my husband learned what his role would be in childbirth and with a newborn. Naima gave him specific tasks: "the mother takes care of what goes in (milk), and you take care of what comes out (diapers)." He had to watch over my hydration during breastfeeding and make sure I had access to berries and nuts so I wouldn't get too hungry. She gave him an active fathering role. Not only did he have to take care of the baby, but he also had to take care of me. Suddenly, I was not the only one responsible for what was going to happen.

According to Naima, this is exactly the kind of hands-on instruction that fathers need during their children's first months of life. After that, they gain confidence in their abilities and continue to take initiative. "Often a couple contacts a doula because the woman wants support, not only for herself, but also material and emotional support for her partner."

"Very often, men who contact a doula want to be involved." Yet, we must not forget that today, calling on the services of a doula (or a) can be expensive. This perinatal service, excluded from private health insurance in the United States and public health system in Quebec, is not accessible to everyone.

Naima also insists that it is not necessary to have had children to develop these very important skills. It is not necessary to wait for men to become fathers to train them, or even that only those who wish to have children get these skills. It's true, there are more and more resources available for fathers. In Quebec, for example, there are many groups, such as La Ligue des Cools Dad, or S.O.S Papa. Elle Quebec has a page dedicated to resources for dads. There are some links, such as How to Bond with Your Child in Utero or How to Make Skin-to-Skin Contact with Your Newborn. Are there other opportunities in the age of AI to receive parental training?

## 3.4 Robot Teachers

While AI, particularly NLP models, can be detrimental to those trying to rethink gender roles around childcare, it has been applied in other ways as well. First, there have been numerous attempts to have AI-powered robots become the childcare providers themselves. Of course, these are quite controversial, particularly from the perspective of parents who are not easily duped by the idea that a robot will be able to provide high-quality care to their toddler. Several scholars have also shown concern that childcare robots would promote antisocial behavior in children.

From my perspective, the idea of leaving a child for several hours alone with a robot seems both dangerous and lonely (for the kid). There is also the question of liability. What happens when a child is harmed by the robot? Companies are keen to evade responsibility, while still selling a product that would help us raise other human beings. One example of such childcare robots is the iPal, developed by AvatarMind. The iPal is a relatively cute-looking robot—the pink one is a girl, and the blue one is a boy. It has a tablet screen at chest level and tells stories through that screen. The difference between the iPal and any old tablet, from my perspective, is that the robot itself is moving around a bit more, which might encourage the child interacting with it to bop around too, probably better for physical activity. Children are much smarter than we give them credit for, however. They won't develop any kind of meaningful relationship with the robot, they will simply try to hack it to get what they want (more cartoons, in the case of my children).

Given the high amount of controversy around robot childcare providers, some companies have turned to another solution—creating AI tools that will help parents be better at their jobs. There are several categories of tools currently on the market for parents and childcare providers. One very common category is monitoring—tracking a child's sleep, activities, grades, location, even harms that might come to them in their digital lives. Another category is sharing—allowing new parents to share photos and other memories from their children, for example, but without the privacy considerations of putting them on Instagram. Finally, there are networking tools, aimed mostly at women, to allow them to meet other moms and socialize.

It can be assumed that each one of these apps uses AI to a certain extent, from facial recognition, to pattern recognition (abnormal activities), to photo or connection recommendations, for example. While certainly less controversial than a robot childcare provider, are these apps doing anything to rebalance gender norms around childcare? Perhaps they could.

## 3.5 Corporate Policies and Social Norms

Diana, the business development manager in Mexico City, is noticing changes in companies, especially in the technology sector. Although women are still in the minority, she has never felt disrespected because of her gender. Although (or perhaps because of) they have also been criticized extensively for gender biases, most large technology companies now offer generous parental benefits, and sometimes childcare in order to retain workers. These policies allowed Diana to be hired a few weeks before she gave birth and to receive paid maternity leave.

In countries where governments are not proactive in promoting women in the workplace, corporate policies play a prominent role. In the

United States, one of the countries with the fewest benefits for families in the world—no maternity or paternity leave, a private and complicated health care system, few protections for workers—technology companies have taken the lead in instituting these types of benefits. Amazon offers 20 weeks of paid leave to biological mothers and six weeks to fathers and adoptive parents. Microsoft offers similar packages, with the addition of support for fertility treatment and adoption. Netflix offers its employees as much parental leave as they need, and according to their website, moms take between 4 and 8 months on average.

These companies grew quickly during the pandemic and had to offer benefits to attract and retain employees. Other companies that have grown, such as consulting companies Deloitte and KPMG, opted to provide similar benefits. However, if the companies or the economy shrink, benefits that are not protected by law can just as easily be changed.

While many couples are successfully pushing the boundaries of gender norms to better accommodate them, government policies that support both mothers and fathers are essential. Indeed, while advances in corporate policies are encouraging signs of change, the benefits they grant obviously do not transcend corporate boundaries. This means that ultimately, only those privileged enough to find a job at a large technology company will benefit.

For his part, Paolo faced systemic challenges more than once when he was a stay-at-home dad. For example, he was teased by his peers who claimed that he was not independent and that being a stay-at-home dad meant he had to ask his wife's permission for everything. When he started taking his oldest daughter to the pool, he was faced with a new issue: how to get her to change into her bathing suit. He ended up encouraging her to enter the women's locker room by herself and wait by the door, biting his nails until she came out. Despite this, Paolo feels that his experience was extremely positive overall.

Other men described the cultural challenges of the job. Leaving early to pick up the kids, sharing pictures of the kids at work, taking a day off if they are sick, taking the kids to the office if you really have no choice. The spillover of family life into work life has been just as taboo for men as it was in the past for women.

It is very difficult for women to develop themselves professionally if men (fathers, politicians, corporate leaders) are not considered as equal partners in raising the next generation. While it is important to continue fighting for policies that will support women, it's also important to acknowledge the need for a cultural shift in relation to men's role in child care. While there are certainly many examples of men taking paternity leave or being self-actualized as primary care givers, this area can still be reinforced. In the age of AI, I haven't experienced very much discussion about how AI might propagate stereotypes about men in care positions. However, any simple search can show that these biases are just as common as biases against women. For example, a test on the DALL-E platform, which generates images based on prompts, has the following effect:

*Prompt:* A loving childcare provider taking care of children.
*Output:* Several images of women smiling and taking care of children.

While having AI replace human caregivers doesn't seem quite right, there may certainly be opportunities to have AI systems provide parental training, help parents to organize tasks, and offer networking and support to caregivers that might be more isolated. Finally, we can also return to the revolution of flexible work, as it allows parents of any gender to be more present with their children and anyone else that they might care for. I recently tried to take my kids to work, and they rolled themselves in the UN flag and almost ripped it, and the 3-year-old then proceeded to sit on his older brother and pummel him. That being said, normalizing children in the workplace, or at least in the background of video calls (for everyone's safety), has been beneficial to men too, as they've also been able to spend more time with their children. Let's just ensure that we set up a structure that supports them as well.

## 3.6 Suggested Discussion Questions

1. What do you think would be the best way of redistributing the responsibility for care work between genders? Could there be different corporate, government, and educational initiatives involved?
2. How do you feel about the lines between work and home, or the public and private spheres, being blurred? Should children walk around in the background of virtual calls, or even be more present in office buildings? Why or why not?
3. Aside from perpetrating more representational images of care providers, what else could AI systems to make care duties more equitable in society?

## References

AvatarMind. (n.d.). *iPalRobot*. https://www.ipalrobot.com/
Brink, K. A., & Wellman, H. M. (2020). Robot teachers for children? Young children trust robots depending on their perceived accuracy and agency. *Developmental Psychology, 56*(7), 1268.
Brusca, R. A. (2017). A comprehensive analysis of the effects of paid parental leave in the US. *Duq. Bus. LJ, 19*, 75.
Choi, S., Yun, S., & Ahn, B. (2020). Implementation of automated baby monitoring: CCBeBe. *Sustainability, 12*(6), 2513.
Dufu, T. (2017). *Drop the ball: Achieving more by doing less*. Macmillan Publishers.
Fournier-Tombs, E., & Castets-Renard, C. (2021). Algorithms and the propagation of gendered cultural norms. Forthcoming for publication in French in "IA, Culture et Médias" (2022). Edited by: Véronique Guèvremont and Colette Brin. Presses de l'université de Laval.
Huang, Q., & Hao, K. (2020, November). The Development of Artificial Intelligence (AI) Algorithms to avoid potential baby sleep hazards in smart buildings. In *Construction Research Congress 2020: Computer Applications* (pp. 278–287). American Society of Civil Engineers.

INEGI. (2010). *Mexico Census of Population and Housing 2010*. https://en.www.inegi.org.mx/programas/ccpv/2010/

INEGI. (2020). *Mexico Census of Population and Housing 2020*. https://en.www.inegi.org.mx/programas/ccpv/2020/

Kaufman, G., & Petts, R. J. (2022). Gendered parental leave policies among Fortune 500 companies. *Community, Work & Family, 25*(5), 603–623.

National Aboriginal Health Organisation. (2004). *Midwifery and aboriginal midwifery in Canada*. https://www.ontariomidwives.ca/sites/default/files/2019-08/NAHO%20Midwifery%20and%20Aboriginal%20Midwifery%20in%20Canada%202004.pdf

Open AI. (2023). GPT-4 Technical Report. *ARXIV*. https://arxiv.org/abs/2303.08774

Pahl, J. (1990). Household spending, personal spending and the control of money in marriage. *Sociology, 24*(1), 119–138. http://www.jstor.org/stable/42854628

Park, R. (2015). What Ruth Bader Ginsburg taught me about being a stay-at-home dad. *The Atlantic*. https://www.theatlantic.com/business/archive/2015/01/what-ruth-bader-ginsburg-taught-me-about-being-a-stay-at-home-dad/384289/

Pollack, S. (2019). 6 Companies Redefining Parental Leave. *MSNBC*. https://www.msnbc.com/know-your-value/6-companies-redefining-parental-leave-n984946

Pew Research Center. (2019). *8 Facts about America Dads*. https://www.pewresearch.org/fact-tank/2019/06/12/fathers-day-facts/

Sagan, A. (2013). Otis Kryzanauskas, Canada's lone male midwife, is an anomaly. *CBC News*. https://www.cbc.ca/news/health/otis-kryzanauskas-canada-s-lone-male-midwife-is-an-anomaly-1.2459342

Sarmiento, I., Paredes-Solís, S., Andersson, N., & Cockcroft, A. (2018). Safe birth and cultural safety in southern Mexico: Study protocol for a randomised controlled trial. *Trials Journal*. https://trialsjournal.biomedcentral.com/articles/10.1186/s13063-018-2712-6

Sharkey, A. J. (2016). Should we welcome robot teachers? *Ethics and Information Technology, 18*, 283–297.

Statistics Canada. (2016). *Changing profile of stay-at-home parents*. https://www150.statcan.gc.ca/n1/pub/11-630-x/11-630-x2016007-eng.htm

Stechyson, N. (2018). Toronto's Ryerson University's 1st male midwife uses his novelty to spread awareness. *The Huffington Post*. https://www.huffpost.com/archive/ca/entry/male-midwife_ca_5cd52ceae4b07bc72975786a

Université du Québec à Trois Rivières. (2018). *À voir: un documentaire sur Louis Maltais, homme sage-femme.* https://neo.uqtr.ca/2018/10/16/a-voir-un-doc umentaire-sur-louis-maltais-homme-sage-femme/

Vogler, C., Lyonette, C., & Wiggins, R. D. (2008). Money, power and spending decisions in intimate relationships. *The sociological review, 56*(1), 117–143.

# 4

# Leadership in the Public Sphere

*Working in a government technology start-up in Monterrey, Mexico, Anita has noticed that the chatbot she is working on, which connects citizens with appropriate government services, seems to have different results for women and for men. It appears to make assumptions based on gender, for example connecting women with family-related services and entry-level job training, while men are advised on financial services and higher-level job opportunities. Anita brings up these errors to her company's management, but they don't take the concerns seriously. There are no women in senior leadership and Anita is not sure how to formulate the concerns. Meanwhile, the government is working on an AI ethics policy which mentions discrimination but doesn't refer to women or gender considerations. Again, the policy development team doesn't include women, and the risks brought forward by Anita are not considered. The start-up management team look at the new ethics policy document but don't see any examples of the kind of errors that Anita is mentioning, so they determine that their chatbot is compliant, and don't address the issues she raises.*

The parity of women in governance, whether in government, boards of directors or companies, has been one of the great issues in gender equality. Women have long struggled to gain access to power, facing

Table 4.1 Current state of women in leadership positions in Canada, the United States and Mexico

|  | Canada | United States | Mexico | World |
|---|---|---|---|---|
| Female CEOs | 15% | 3% | 8% | 23% |
| Women in C-suite positions | 31.50% | 34% | 26% | 29% |
| Women in board seats | 27.60% | 21% | 6% | 10% |
| Women in parliament | 29.60% | 26.50% | 48.20% | 25.20% |
| Female cabinet members | 50% | 48% | 42% | NA |

barriers, ceilings, and all kinds of other structural blockages. In many societies, the lack of women in leadership positions is glaring. In many countries, the percentage of women in board positions is in the single digits, particularly in technical sectors, such as energy. At the same time, there is a surge of support for women and leadership, particularly on boards of directors—where countries not supporting this are seen as outliers. Nevertheless, no woman has been elected Prime Minister of Canada, President of the United States, or even Secretary General of the United Nations. Women's leadership in society is a struggle that is far from over. If it were, we'd need to see on average 50% of governance positions in various sectors held by women—still a work in progress.

The norms that lead to these inequalities are still difficult to eliminate. On the one hand, there is the hierarchical system that still in many areas overvalues the contribution of men. On the other hand, there are unfortunately still toxic perceptions of women in power, especially those related to their competence. Finally, as we saw in the last chapters, there is also the volume of domestic work that many women still perform, making their access to power even more complicated (Table 4.1).

## 4.1 Barriers to Women's Leadership at Work

Some time ago, I was invited to speak at a conference for students at a university in England. An extremely well-prepared young woman asked me if I had ever experienced discrimination in my workplace. "Of course," I said, but I wasn't sure how to elaborate. Indeed, when a woman is not hired for a position or promoted, she is not told if it is because of her gender, or if it is simply a question of her competence for the position. During my two maternity leaves, a more senior position opened up for which I was not considered. The first time, another woman was chosen, while the second time, it was a man. I couldn't tell exactly if that was sexism or bad timing or maybe I was just not the right person for the job. When it comes to sexual harassment in the workplace, which almost every woman I know has experienced, things are a little clearer. On the other hand, it is impossible for me to know what my personal trajectory would have been if I had been a man. Has being a woman an advantage? It certainly has in terms of my credibility to write a book about women and technology. Has it been a disadvantage? Perhaps, but I can't quite tell.

However, when I spoke to women in the context of this book, they all told me about frustrations that were similar to mine. Well-educated, they started their careers well with interesting internships and junior opportunities, until a point in their thirties when they were ready to move on to the next level. That's when they felt held back (did they hit the infamous glass ceiling?) and began to see their male peers surpass them.

Jane, for example, held several high-level positions in government before leaving to become a consultant. Today, she plans to develop a tourism business that she can run from her home. "I was 22 and full of ideals, like most graduate students" she says. "I was single, working very long hours, but I didn't mind because I loved my job. Yet when it came to promotions, only single women were asked. The supervisors knew that the hours would be very long, very demanding, and they implicitly thought that married women would not be able to do the job, because they had more limitations and challenges."

When Jane herself finally left her job, it was not because she had children, but rather because she thought she would never be able to start a family in this context. This was a classic example of the family or career dilemma, a choice that many women still feel they must make. Yet women in many countries have the right to maternity leave, and access to daycare and, in many cases, housekeepers and nannies. Yet Jane felt double pressure—from her supervisors to remain single, and from her family to slow down and start a family.

While discrimination is made illegal internationally by CEDAW, it is nonetheless still common. The way we work now, with long hours sacrificed at the office (or at home), is a relic of another time, and proven to be much less productive than it appears.

## 4.2 The Trap of Long Work Hours

In a previous job, I also had to stay late at a client's office several times without knowing when I would be able to go home. I had a 7-month-old baby, and I worried a lot about leaving around 9 p.m. to nurse my son one last time before he went to sleep, fearing that my career advancement would be jeopardized. During these times, only colleagues who did not have additional responsibilities could continue to work until midnight. While Jane is correct that those with outside responsibilities could not manage the work schedule, the problem was not so much with the mothers as with the organizational culture. What's so urgent that a government administrator or even a consultant must regularly stay at work late at night, without warning? Indeed, according to Parkinson's Law, the time required to complete a task increases in relation to the time allotted.

One of the ways of measuring productivity is the OECD's GDP per hour of work calculation, shown below for Canada, Mexico, and the United States. There are certainly flaws to this approach, as there are many factors to a high GDP, not only productivity. In this table, the United States would be the most productive, and as the country with the largest economy in the world, it is considered to also have the most productive workforce. It would also indicate that number of hours worked is not necessarily indicative of a high economic output (Table 4.2).

There may be other, better, ways of showing this; however, there is still a paucity of research in terms of worker productivity and hours worked. An interesting study by Marta Angelici and Paola Profeta noted that, in a large Italian company, flexible work arrangements significantly improved productivity and also allowed fathers to take more part in care activities at

Table 4.2 Average hours worked and GDP

| Country | Average hours worked in 2019 | GDP per hour of work |
|---|---|---|
| Canada | 1670 | 58.3 |
| Mexico | 2137 | 22.2 |
| United States | 1799 | 74.8 |
| OECD | 1726 | 58.4 |

home. These results were obtained by comparing the outputs of a group that had flexible work hours with no constraints, to a control group.

In my experience, those who stayed late at the office were not working on urgent matters that could not be postponed; rather, they were trying to signal that they would work harder than others—to survive in a very competitive environment. It doesn't have to be this way. I have also experienced an office where everyone tried to be efficient in the time allotted, and we were no less productive for it.

Having outside responsibilities requires better planning and more efficient use of time. Many studies show that most workers are productive for three or four hours a day, well below the traditional 9 to 5. Women like Jane should have been able to easily accomplish their tasks within the allotted time, after which they, like their male colleagues, should have been free to go home and spend time with their families or whomever they wished. The solution to this problem is not to continue to reduce women's promotions and job responsibilities. Instead, we should focus on rebalancing household duties and better managing work.

In fact, many other studies have shown that productivity does not increase linearly with the number of hours worked. An English study has shown that office workers can accomplish as much in less than three hours as in eight hours. An experiment conducted by Microsoft Japan in 2019 showed that 4-day weeks increased productivity by 40%.

In fact, for knowledge work in particular, conference calls and appointments are considered the most important hindrance to productivity. Employees who work long hours are therefore not more productive, but spend long hours in conferences that could be shortened without affecting work, quite the contrary.

## 4.3 Negative Perceptions of Female Leaders

When it comes to power, there are significant gaps in corporate leadership and governance in almost every sector. So much so, in fact, that studies have shown that many people are not convinced that women can be competent leaders. In a 2007 report, the Catalyst Research Institute described a double bind for women in the workplace, where women must achieve much higher levels of competence than men to be respected.

I spoke recently to some of the founding members of Voz Experta. This organization seeks to correct gender imbalances in the Mexican energy sector. Created by a group of senior women in the sector, Voz Experta specifically targets expert groups. In the energy sector, as in many others, companies and professional associations regularly organize conferences on topics of interest. Voz Experta is an association of about 250 experts—finance specialists, lawyers, electrical engineers, geologists, and other professional women who have at least 10 years of experience in the field. Every time that the organization sees a discussion panel on energy that has no women—she calls them a "Manel," one of the 250 experts on the list is offered in order to add balance and diversity. In the 2.5 years since it was created, the organization has been able to influence close to 100 panels.

In Mexico, there are hundreds of these conferences every year, on topics such as sustainability in the oil and gas industry, electricity privatization, or solar panel installation. Although there are many capable women working in this sector, they are underrepresented at these conferences, which are often all-male. So, the women of Voz Experta called the organizers of more than 200 conferences in recent years to encourage them to include women experts.

"It's funny that many men don't perceive that there's a problem," says a co-founder. "When I talk to my friends or we meet with allies, they are shocked to see pictures of these meetings where there are only men. They can't believe it because they think they are feminists and they support women. Just making that change and raising awareness about it is really good."

I've talked to a number of women about this organization because it fascinates me. The problem of gender equality is very complex, yet the solution this group offers is simple and effective. Faced with one of the worst gender imbalances in any industry (only 3% of Mexico's energy board members are women), they address the perception that women are not competent in the field by allowing them to present their knowledge in public. When they approach the organizers of these conferences and ask why they haven't invited any women, they are invariably told that there are none to invite. Like AI algorithms that blindly repeat and reinforce gender stereotypes, conference organizers make reckless assumptions and thus contribute to the perception that women are not competent enough to have a voice.

In 2021, the UN office in Indonesia published a manual to avoid conferences without female representation, or "manels." According to the document, "Manels perpetuate the underrepresentation and misrepresentation of women by completely ignoring them." The office considers this problem, exemplified by Voz Experta in Mexico, to be a global issue for women. It therefore proposes eight steps, such as the inclusion of women in the organization of conferences, and the revision of participation criteria, to encourage speakers to invite women to speak.

However, many ambitious women feel frustration when it comes to their career advancement, and may not want to wait to be invited to expert panels. Another solution, for better or for worse, is the "third way," which consists of leaving the traditional work structure completely and working as an independent consultant. Flexibility, freedom, and no glass ceiling, what's not to love? (I will propose that in this construct, the "first way" is staying home, the "second way" is going to a salaried job, and the "third way" is being a freelance consultant).

## 4.4   The Third Way: Consulting

I recently spoke with Erin Halper, CEO and founder of The Upside, a company that supports consultants who work with companies on what she calls "the four pillars of value": revenue growth, cost savings, access and networking, and problem-solving. Halper, like many of her clients, left a high-paying job in the private equity industry to have more flexibility. When she found out that her first son needed a series of surgeries, that flexibility meant she was no longer "chained to her desk," but could continue to work while caring for him in the hospital.

"I was able to happily work on a laptop from his hospital bed, and it wasn't a burden," she says. She experienced similar freedom after the birth of her second child, when: "after a few months, I couldn't wait to get back to work, because I had full control over how, where and when I worked."

For many women, returning to work is not so simple. "They come from a full-time corporate culture where there were micro-abuses over their 10–15 year careers, where they were made to feel like they were lucky to have this opportunity, and there's a line of other people who want their jobs. So, they accept lower wages and working conditions because they have to be grateful. They leave the company with a kind of abused mindset. But in consulting, it's 'you're lucky to have me and my work contributes a lot'."

The statistics on women in corporate leadership are staggering. In 2019, women held 18.2 percent of corporate board seats in Canada, and only 3.5 percent of companies had a female CEOs. The numbers are similar in the United States, where between 20 and 25 percent of board seats were held by women in 2019, depending on the sector. In Mexico, 6% of board seats were held by women in 2018, with the highest proportion in the consumer goods sector (36%) and the lowest in the energy sector (3%).

While these statistics are very useful in quantifying discrimination against women and allowing us to track progress, they are also a bit misleading. Although women have a much harder time climbing the corporate ladder than their male counterparts, that doesn't mean they aren't finding other ways to have a fulfilling career.

Many of the women I spoke to left their corporate jobs to become consultants—some because they wanted to spend more time with their children, others because they were frustrated by the discrimination they faced. They felt that consulting gave them the independence to choose their clients, work at a higher level of responsibility, and create their own schedules.

In my case, I started working as a consultant after I returned to Montreal, when my oldest son turned 2. Over the next four years, I became increasingly selective in the work I chose, developing a portfolio of mandates as an expert in my field. During this time, I worked harder than ever, also experiencing much more motivation and job satisfaction than before. I did find this period of time quite interesting. In the same week, I could write a report on climate change for an international bank, collaborate with a colleague on a scientific paper on new research methods, give a talk on the digital rights of refugees, and train a team of 20 data science volunteers. I attributed this productivity to, among other things, the fact that I now had much more flexibility and autonomy.

Hearing Halper speak so passionately about the amazing work her clients do as executive trainers, brand strategists, and technology experts made me realize that women have developed an alternative structure, outside of traditional companies, that is addressing power imbalances in the corporate world.

In 2016, MBO Partners, a US-based company that helps companies work with independent consultants, found that women made up a slight majority (but a majority nonetheless) of freelancers in the United States. This trend is also found in Mexico and Canada, where women find more autonomy, empowerment, and even financial compensation in the consulting field.

However, while many women have found an advantage to being self-employed, they also have a more precarious status. Financial security is thus sacrificed for autonomy, something that can certainly impact their income and financial independence if something goes wrong.

During the pandemic, for example, women were most economically affected by confinement and their poverty level increased globally. According to UN Women and UNDP, the poverty gap between women and men was widened by COVID-19. In an Oxfam report, the authors also note an increase in unpaid care work and financial instability among women since the start of the pandemic.

The third way is thus as much a symptom of discrimination against women as it is a solution. In the current structure of work, which can be harmful and complicated for many women, it is natural that many women turn to alternatives. However, in the long-run, women are also much more prone to take part-time and unstable work.

## 4.5 Women and Motherhood in Government

While there are many barriers to women being perceived as competent and holding positions of responsibility, a number of news articles have recently argued that women are in fact more competent leaders than men. In particular, articles such as "Women Were Better Leaders Than Men During the Pandemic" assert that leaders such as Tsai-Ing Wen of Thailand, Jacinda Ardern of New Zealand, and Angela Merkel of Germany were undoubtedly more successful in containing the crisis in their countries than Jair Bolsonaro in Brazil and Donald Trump in the United States. However, this ignores circumstantial contexts and cultural parameters. Moreover, making claims about a gender's superiority and inferiority is contrary to the aims of gender equality.

I do think that articles like these that may help change stereotypes about women in leadership roles. However, such talk can also be detrimental to women, arguing that they deserve leadership because they work harder than men, rather than simply because they are human beings, and all human beings have the right to seek leadership if they so wish.

This was stated very succinctly by a senior corporate leader, who stated: "as a woman, you have to do four times the work for a third of the profit. Every woman in a leadership position that I interviewed said more or less the same thing. In today's context, it is realistic to assume that women in power have had to work much harder and therefore have achieved higher levels of competence in the process."

This is why I am uncomfortable when I hear leaders say that women are "unquestionably better" than men. Is this the justification for women's equality in the public sphere? Does this mean that we only deserve equal representation in corporate leadership and governance if we are better than men? In reality, it continues to perpetuate the notion that women are inherently less valuable than men; only if they perform far better than men do they deserve equal power.

Pascale Navarro, author of the book *Are Women in Politics Changing the World?* asked dozens of Canadian women in politics why we should strive for parity. They all had the same answer: in a democratic system, elected officials should be representative of the citizens. Thus, when at least half of the population is female, as in all countries, this should be reflected in the political sphere. This equality of representation applies to people of color, Aboriginal people, transgendered people, and any other group that makes up a given nation.

Organizational psychologist Nancy Doyle has also argued that this current wave of gendered leadership traits is condescending and reductive. There are many reasons why Jacinda Arden might have done better in New Zealand, not the least of which is her gender. Beyond her skills, which nonetheless seem excellent, New Zealand is an "isolated island with a small population." Other islands, such as Jamaica, have also done better than countries with many land borders. Dr. Doyle quotes neuroscience professor Gina Rippon, who believes that "reinforcing gender differences is socially artificial and may actually set women back, rather than advancing the acceptance of female leadership."

Finally, there is the issue of the place of motherhood in women's leadership. Being a mother not only brings special responsibilities (such as childbirth and breastfeeding), but it also provides an opportunity for female politicians to better understand this important part of the electorate.

However, in 2007, Dana Goldstein strongly criticized women politicians for openly "pushing" the fact that they are mothers during their campaigns. Why, she asked, should women have more professional value as mothers than simply as competent human beings? At the time, a number of feminist writers were concerned. After all, positioning herself as an advocate for women might give a politician more votes from the group she was advocating for, but would she then lose her broader appeal as a representative of all?

Goldstein continued: "Here was the highest-ranking woman to ever hold elected office in the United States, talking proudly about breaking the 'marble ceiling' of the U.S. Capitol, but flooding the dais with children and bragging about her journey from 'the kitchen to Congress,' smiling blissfully as she cradled her newborn grandson."

Certainly, women politicians emphasized that they were mothers because they felt that this gave them added value as leaders. They were able to represent the perspective of motherhood, which had been underrepresented in government. Did they feel they had a better chance of being elected by running for office with an understanding of and ability to advocate for mothers' issues?

According to a study published in the American Political Science Review, the answer is yes. Indeed, the authors note that, "Overall, elites and voters prefer candidates with a traditional family profile, such as being married and having children, which results in a double bind for many women." This is further complicated by the fact that many female politicians, such as Angela Merkel in Germany or Kamala Harris in the United States, do not have children, but still have to project a soothing, maternal image.

Certain commentators feared that this political card might label these leaders as competent only in women's issues. This could call into question their ability to govern in the broader public interest with voters. But this is perhaps because until very recently, motherhood was not seen as an asset in leadership and public life. The division between the private and the public sphere meant that making the private public would undermine a woman's credibility. This perspective has begun to change, however, with the broader acceptance of two ideas—first that motherhood, or any care work, can be seen as a very skill-intensive activity—skills that are certainly transferable to other domains (organization, patience, resilience, problem-solving, communications, leadership, diplomacy, self-control, and I could keep going—did I mention resilience?). And second, that women have a responsibility to bring forward issues that have hitherto not been of public concern. So, for now at least, women in power have the dual responsibility of representing all constituents and raising the issues that affect women specifically.

Given the current cultural context, it is both true that women must advocate for issues that affect their gender, children and the elderly (whom they often care for), and that they can lead with a broader perspective. The presence of women in leadership positions allows our society to draw on a greater number of capable leaders, in addition to offering a diversity of perspectives. This change opens up new perspectives that could help rebalance gender norms over the long term.

Against this backdrop, it is unsurprising that women have had less opportunities for leadership in the AI sector. Women's leadership in the public sphere, whether in business, technology, or politics, is critical in the age of AI. AI is a sector with immense dynamism and growth, while also presenting significant threats if it is not managed properly. It represents therefore a field where female leaders can not only seize opportunities for significant positive impact, but also prevent future harms to gender equality. Simply, women, and any other human being, if they so choose, should have the opportunity to influence the society in which they live. Today, that is the field of AI. Tomorrow, it will almost certainly be another battleground, and we will lead there too[1].

## 4.6 Suggested Discussion Questions

1. Do you consider the "third way" a good path for women seeking to break the glass ceiling? What are certain factors that might affect the desirability of this path?
2. Do you agree that equality in leadership might mean that all genders could make both bad and good leaders? Why or why not?
3. Should leaders talk about their parenting or care duties as assets? What if we achieved full gender equality in leadership, would leaders speak more about their children, or less?

## Note

1. Article par Kamala Harris publié en 2019 par le périodique Elle Magazine: https://www.elle.com/culture/career-politics/a27422434/kamala-harris-stepmom-mothers-day/

# References

Angelici, M., & Profetea, P. (2023). Smart working: Work flexibility without constraints. *Management Science*.

Catalyst. (2018). *The double-bind dilemma for women in leadership: Damned if you do, doomed if you don't (Report)*. https://www.catalyst.org/research/the-double-bind-dilemma-for-women-in-leadership-damned-if-you-do-doomed-if-you-dont/

Catalyst. (2020). *Women in the workforce in Canada*. https://www.catalyst.org/research/women-in-the-workforce-canada/

Catalyst. (n.d.). *Women in Management*. https://www.catalyst.org/research/women-in-management/

Champoux-Paillé, L., & Croteau, A.-M. (2020). Pandémie: les femmes Font-elles de meilleurs leaders? *The Conversation*. https://theconversation.com/pandemie-les-femmes-font-elles-de-meilleures-leaders-137048

CTV. (2019). *Regional and gender breakdown of Justin Trudeau's new cabinet*. https://www.ctvnews.ca/politics/regional-and-gender-breakdown-of-justin-trudeau-s-new-cabinet-1.4696687

Doyle, N. (2020). Do women make better leaders during a pandemic? Don't trust the data… *Forbes*. https://www.forbes.com/sites/drnancydoyle/2020/08/20/do-women-make-better-leaders-in-a-pandemic-dont-trust-the-data/#18c48dbd3073

El Financiero. (2021). *Paridad de género al gabinete llegó con Gobierno de AMLO: Sánchez Cordero*. https://www.elfinanciero.com.mx/nacional/paridad-de-genero-al-gabinete-llego-con-gobierno-de-amlo-sanchez-cordero

Harris, K. (2019). Sen. Kamala Harris on Being 'Momala'. *Elle Magazine*. https://www.elle.com/culture/career-politics/a27422434/kamala-harris-stepmom-mothers-day/

Goldstein, D. (2007). The mommy mantra. *American Prospect*. https://prospect.org/article/mommy-mantra/

Gupta, A. (2021). Fulfilling a promise: A cabinet that 'looks like America.' *New York Times*. https://www.nytimes.com/2021/01/21/us/biden-cabinet-diversity-gender-race.html

IPU. (2023). *Monthly ranking of women in national parliaments*. https://data.ipu.org/women-ranking?month=1&year=2023

McKinsey. (2018). *One aspiration, two realities*. https://womenmattermx.com/en/WM_Nov_ENG_final.pdf

Mercer. (2020). *Let's get real about inequality.* https://www.mercer.com/con
tent/dam/mercer/attachments/private/gl-2020-wwt-global-research-report-
2020.pdf

Navarro, P. (2020). *Les femmes en politique changent-elles le monde?* Boréal.

OECD. (2023a). Hours worked. https://data.oecd.org/emp/hours-worked.htm

OECD. (2023b). *Labour productivity levels: Most recent years.* https://stats.oecd.
org/index.aspx?queryid=54563

OSLER. (2021). *Report: 2021 diversity disclosure practices — diversity and leadership at Canadian public companies.* https://www.osler.com/en/resour
ces/governance/2021/report-2021-diversity-disclosure-practices-diversity-
and-leadership-at-canadian-public-companies

OXFAM Québec. (2020). *COVID-19: Augmentation du travail de soins non rémunéré et de la détresse des femmes.* https://oxfam.qc.ca/covid19-augmentat
ion-detresse-psychologique-femmes/

Spencer Stuart. (2019). *Mexico board index 2018.* https://www.spencerstuart.
com/-/media/2019/may/mexico_board_index_2018.pdf

UNDP. (2020). *COVID-19 will widen poverty gap between women and men, new UN women and UNDP data shows.* https://www.undp.org/press-rel
eases/covid-19-will-widen-poverty-gap-between-women-and-men-new-un-
women-and-undp-data-shows

United Nations Indonesia. (2021). *Guidance for avoiding all-male panels.* https://indonesia.un.org/sites/default/files/2021-06/Guidance_No%20M
anel_0.pdf

Teele, D., Kalla, J., & Rosenbluth, F. (2018). The ties that double bind: Social roles and women's underrepresentation in politics. *American Political Science Review, 112*(3), 525–541. https://doi.org/10.1017/S0003055418000217

Twitter (n.d.). Voz Experta. https://twitter.com/vozexpertamx?lang=en

World Bank. (2022). *Proportion of seats held by women in national parliaments.* https://data.worldbank.org/indicator/SG.GEN.PARL.ZS

# Part II
## Being

This second section is about bodies and the self in the age of AI. The three chapters discuss perceptions of the body and physicality, and physical security. It is difficult to write about perceptions of women at work and in positions of leadership without writing about stereotypes related to their bodies. While in the last section we discussed the way in which historical context influences AI systems to perpetuate gender stereotypes in the work place, in this one we discuss sexualization and misogynistic speech, which conspire to discredit women's contributions and to threaten their physical safety.

There are so many examples in which women in positions of leadership are asked questions referring to their bodies, seemingly reducing the value of their intellectual contributions. From fashion-related questions to prime ministers and heads of state, to comments disparaging female leaders because of their weight or physical attributes, the objectification of women continues online and offline, effectively threatening their ability to contribute to society on an equal footing.

The impact of misogynistic speech, which appears to have increased on social media during the COVID-19 pandemic, is devastating to women's security. This type of speech, which is spread easily through social media's polarizing algorithms, has been linked to emotional and psychological damage, as well as offline physical actions such as domestic violence and sexual assault. Largely, this takes place because new technologies, including social media platforms, are not sufficiently analyzed through a Women, Peace, and Security (WPS) lens. By this, I mean that when we launch a new product and determine that it's okay to "launch first, ask questions later," we don't consider the real-life effects to women's security of having ill-conceived and ill-monitored systems.

In the age of AI, we certainly have the opportunity to use new technologies to address issues affecting women's bodies, from sexualization, to violence, to strength. However, this must be done intentionally, with more efforts toward fixing stereotypes in generative AI, reducing the amount of hate speech on social media platforms, and investing in applications that could support women's empowerment. Fundamentally, we need to care enough about women's safety to prioritize this when designing AI.

In the next three chapters, we will dive deeper into these issues—examining stereotypes around the physical representation and athletic ability of women, exploring threats and opportunities to women's peace and security online, and better understanding how gender stereotypes can be embedded into AI systems.

# 5

## Strength and the Power of Sport

*Michelle is sitting on the sofa next to her boyfriend, Mike. Each one has a laptop open on their lap and is browsing through Netflix. Michelle's screen shows romantic comedies and reality TV shows about fashion. Bored with the selection, she peeks over at Mike, who has a section on sports, including a weightlifting documentary. Curious, Michelle types the name of the documentary in the search bar, and retrieves it. Working from home, accessing entertainment mostly through news and video recommendation systems, she doesn't know anything about strength sports. However, in this documentary, she learns about Olympic weightlifting, and sees many women practicing this sport. Over the course of the next few weeks, she continues to look at the content suggested to her boyfriend, and realises that it is completely different from hers, and that, sitting side by side on the couch, they inhabit totally different worlds.*

Shortly after we moved into a tiny apartment in Brooklyn, I accompanied Curtis to the South Brooklyn Weightlifting Club, a black iron gym where followers practice powerlifting and weightlifting. He had been doing strength sports since he was a teenager and had heard about these types of gyms from a friend of his at work. The gym was a short walk from our house and next to a friendly pub we were going to visit afterward. As I entered what looked like a small warehouse, I was struck by the seriousness of the men and women loading and unloading weights onto their respective bars. A few people were chatting near a table that held a bowl of gummy bears, munching as they caught their breath. The owner, a round, friendly man named Paulie, walked up to us with a smile.

While Curtis took a quick tour of the gym, I stood in the hallway, staring blankly at my phone and feeling a little out of place. At the time, while I loved running and yoga, I had never touched a barbell, and I was mesmerized by the quietness of the room, punctuated by the sound of weights hitting the floor. A group of women were shouting encouragement to each other: "Knees out! Come on! You can do it!".

Paulie approached me. "You know, you can sign up for a couples package with Curtis, you can train together!" Always up for a challenge, I quickly agreed. And so began my five-year journey into strength sports, where I competed at the grassroots level with Curtis, got really strong, and got to know a new kind of feminist, the strongwoman.

In fact, these types of gyms are at the forefront of a progressive women's movement in sports. Women are taught to focus on strength and muscle rather than leanness. It was therefore a shock for me, at the beginning of my training, that I was asked to gain weight. "You have to eat a donut a day," the coach advised me. To be completely honest, I didn't do it, but the strongest women in the gym all ate well. To me, encouraging a woman to gain weight was unheard of! Instead of daily donuts, I ate Greek yogurt and peanut butter, and gained 22 pounds in a year. Very quickly, I started doing local competitions. A little too tall compared to my weight, I was not the strongest, but I managed to squat one and a half times my body weight, and push 50 kilos on the bench press. Not bad for a beginner! What surprised me was that two of my friends had also independently taken up powerlifting and had the same, if not better, results. Both quite lean, having never touched a workout bar in their lives, they too were able to build muscle quite quickly.

Powerlifting is probably not the most well-known strength sport. However, the atmosphere of the gyms dedicated to it is magical. More and more women practice this sport and succeed very well. In Montreal, there was, until 2019, a gym in Griffintown dedicated to this sport. The owner, a young woman from the small town of Saint-Lambert, encouraged women a lot. I saw that same atmosphere in Brooklyn, where I practiced the sport for three years. It was an atmosphere of empowerment, self-actualization, and also leadership.

In my opinion, the place of women in strength sports is linked to the place of women in the world of work. Why? Because in many fields of work—including leadership, technology, and science, among others—women have often been told that they are incompetent, unable to reach the same level as men. Likewise, there have been many barriers to the inclusion of women in strength sports. We were often told that we couldn't lift as much as the men, that it wasn't worth trying. Yet the ability to practice a sport, just like leadership, is an extraordinary asset in our society, although it is undoubtedly undervalued. There is no reason for women to be deprived of the possibilities and well-being that physical power offers.

## 5.1 Women in Strength

There are several disciplines within strength sports, including not only powerlifting, but also strongman and weightlifting. The latter was included in the Olympic Games in 1896. However, it was not until those of Sydney in Australia, in 2000, that a female category was created.

This year marks an important step for this sport which is seeing an increase in its number of entries. Between 2007 and 2018, the percentage of women registered with USA Weightlifting, the American organization that organizes weightlifting competitions, increased from 19 to 37%. Many countries have established very competitive women's weightlifting teams, including Canada, Mexico, and the United States. Today, nearly half of the athletes on the Olympic weightlifting teams in the United States and Japan are women.

This popularity does not prevent female weightlifters from being confronted with many prejudices. This sport encourages the development of physical strength, hitherto reserved for men. Although this is starting to change, women's musculature is still often associated with something abnormal, too far from the stereotypes of female beauty sold in magazines.

Diana Furhman was one of the pioneers of female weightlifting in the United States. In the 1980s and 1990s, she took part in national and international competitions. Her coach, Bob Takano, explained at the time that female athletes were trained differently than their male counterparts. One of the obstacles to Furhman's full potential, he conceded, was that she wanted to stay thin. She long feared going against social norms and gaining muscle mass. Her competitiveness for her weight class was questioned, and it wasn't until she let go of the pressure to be thin that she really excelled. She won several championships and set the American standards of her generation.

Athletes like Furhman, however, who manage to overcome the barriers of norms and surpass themselves, are role models for many women and young girls. For my part, I particularly admire Heather Connor, a tiny woman of 47 kilos, who won the powerlifting world championships in 2019. At 29, she had been training for 4 years while continuing her day job as a kindergarten teacher in North Carolina. What impresses me is that she managed to squat almost triple her weight—136 kilos—and deadlift even more. By way of comparison, power lifter Thor Bjornsson, the Icelandic who played the character *The Mountain* in the *Game of Thrones series* and considered one of the strongest men in the world, squatted 445 kilos for a weight of 205 kilos, a ratio of only 2.2 times his weight.

While social norms related to strength are changing, they still exist. When I was training a lot, how many times did I hear that I wouldn't be able to fit through the door frame, or that I should be careful not to get "too bulky." Well-meaning men and women warned me that men didn't like women that were too strong. Others taunted me saying that I shouldn't bother, since I would never be as strong as a man.

This kind of discrimination in weight rooms, often reserved for men practicing strength sports, is well known. A group of researchers from the University of Waterloo in Ontario found that women often find it difficult to train in weight rooms, feeling more comfortable near cardio machines. The low representation of women in these sections of the gyms is undoubtedly a cause, but there are certain elements of the culture that can make it downright hostile for a woman. For example, a gym in the UK was heavily criticized for using an image of a pair of female buttocks in a string bikini in its advertisements. This kind of image signals to women that they are there not to participate in the sport, but to be objectified. This can make the experience very uncomfortable for those who are interested in pursuing a strength sport.

I met Brittany Guillory at the black iron gym in Brooklyn. She is a teacher and has competed in weightlifting and powerlifting for several years. She has also worked as a personal trainer, focusing on promoting strength for all women, and especially black women. During one of our exchanges, she said to me: *"I knew I was strong emotionally and mentally, and I wanted to be strong physically. It's one of the many things I had to unlearn, that women don't have to be thin."*

Like many weightlifters, she faced resistance from those around her. *"My family and the community around me did not appreciate the strength of women. They did not understand that it is not a threat if a woman is physically strong. There were lots of sports opportunities for boys and young men in my community, but only things like ballet and theater for girls."* As an aside, I myself have done ballet and I can say that ballerinas are *extremely* strong, as are gymnasts and many others who practice traditionally female sports. However, for having lived it, dancers are under a lot of pressure to appear slender and delicate.

During my journey in the world of strength, I met many women who, like Brittany, rejected these physical stereotypes. When I embarked on this path, I became more aware of the enormous pressure that women are under to conform to certain body types. It takes a tremendous amount of time and energy to change our bodies to what society demands, time that could be better spent on leadership and social impact.

Weightlifting and powerlifting are not the only sports that reflect unequal gender dynamics in our society. Moreover, sport is one of the professional fields where we see the most gender inequalities, particularly in terms of opportunities and salary.

## 5.2 The Superwoman Schema

The backlash against women and strength has not necessarily been a straightforward one. In fact, while there has been pressure for certain women to present themselves as weak, many women have had a reverse experience, feeling that any vulnerability on their part has been discredited.

Wanting to further explore the linkages between strength sports, women, and race, Brittany Guillory surveyed several fellow African American, female lifters about their experiences for this book. Several themes came up in the interviews—the issue of representation in the sport, the recurrence of discrimination, but also the empowering experience of lifting, and the importance of physical, intellectual, and psychological strength.

One woman succinctly described the contradictions in the way she felt strength was considered for women: "*In strength sports, at times, I was expected to be strong because I'm Black. Then, I was made out to be weak because I am a woman. I was berated and belittled until I could lift three times my own bodyweight. I had to outperform just to be accepted. But then when I outperformed, it wasn't surprising because of my ethnicity. To this day, I am still confused.*"

At the same time, African American athletes such as Simone Biles and Serena Williams have faced enormous amounts of criticism for appearing to be too strong. Because they are at the intersection of two discriminated against groups, they are considered both too strong and not strong enough.

One outcome of this contradiction in norms has been the treatment of Black women in health care in the United States. African American women are much more likely to die in childbirth, a fact that has been attributed at least in part to the fact that their opinions on their own health are not legitimized by doctors. In the United States, this combination of systemic racism and sexism has contributed to maternal mortality rates in Black, American Indian and Alaskan Native women being almost five times higher than in white women.

In Quebec, the Aboriginal woman and mother Joyce Echaquan died in the hospital in October 2020, after having been berated on camera by attending nurses. She was administered medication that she was allergic to, and tried to communicate this to the medical staff, who countered with dismissive and racist slurs. Hours later, she was dead.

There are so many more examples of women facing discrimination in relation to their bodies. They are not respected when it comes to how strong they want to be, how they want their body to look, or what kind of medical care they need. The feminist writer Roxanne Gay, writing about women and body image, has captured what many women have felt about society's interactions with their bodies: "*People see bodies like mine and make their assumptions. They think they know the why of my body. They do not.*"

## 5.3 The Sports Pay Gap

The pay gap between men and women is glaring in professional sports. For example, since 2020, Naomi Osaka and Serena Williams, the two tennis champions, are the only two women to appear on the list of the 100 highest paid athletes worldwide. This is, however, the first time that there has been more than one woman on this list. The third highest paid woman, tennis player Ashleigh Barty, earned $13.1 million in 2020, half of what the 100th athlete on the list earned—French soccer player Antoine Griezmann.

However, in the United States, female tennis players are much closer to pay parity than their peers in other fields. While their salary is 84% of that of a male tennis player, it's a lot compared to other sports. In baseball, women earn 0.15% of men's wages. NBA players are also estimated to earn 100 times more than WNBA players. In Mexico, the highest paid female professional soccer player earns less than $19,000 per year, while her male counterpart earns $4.8 million per year. Even more troubling, more than 90% of female players in the Mexican national soccer league earn less than $316 a month, despite the success of their sport. Finally, in Canada, the gaps are similar. Women in the Canadian Hockey League earn between $3,300 and $12,000 each per season, with a maximum of $300,000 in salaries per team. In 2019, the players went on strike, demanding fair compensation. Amanda Kessler, one of the best-known hockey players in the league, earned around $10,000 in 2019. Her brother, Phil Kessler, who plays in the NHL, earned nearly $10 million in the same year.

There seem to be two main reasons for these discrepancies. First, UNESCO reports that women's sports receive 4% of airtime globally, compared to 96% for men's sports. Advertisers have therefore argued that the investment in women's teams is not worthwhile, since it will not pay off. A claim, that, of course, is illogical, since women are also active consumers. Moreover, this argument falls apart in the case of the National Women's Basketball League, women get 81.6% of the overall airtime, and still receive only a fraction of the salary of their male counterparts. However, inequality in terms of airtime is significant because it perpetuates a low representation of women in sports. This is also accentuated by a lack of sponsorship contracts for women, who find it difficult to find alternatives to increase their salary.

## 5.4 Sport for Equality

On the other hand, sports participation for women is important on many fronts. Numerous studies show that sport is beneficial not only for the individual who practices it, but also for improving gender equality.

A study by professional consultancy firm EY found that although women make up just 3.4% of CEOs globally, 94% of female CEOs play at least one sport. According to the researchers, participating in sports allows girls and women to develop not only their work ethic and determination, but also their competitive spirit, which may allow them to be more successful at work. A similar study by the *Peterson Institute for International Economics*, in Washington, DC, finds that girls who play sports have better grades in school, fewer health problems, and higher earnings as adults.

Many community organizations use sport as a tool to increase gender equality. However, in a survey of 10,000 girls in Canada in 2020, Canadian Women in Sport found that two-thirds of girls do not play any sport. The percentage of women over the age of 15 who practiced sports has dropped dramatically in the past decade. It rose from 50% in 1992 to 35% in 2010 and then to 18% in 2020. Like previous research, the study nevertheless found many benefits to women and girls' participation in sports, including well-being, physical and mental health, body image, leadership, and self-confidence.

Globally, UNESCO is leading the charge to rebalance sports participation, remuneration, and airtime. In 2015, the organization adopted the International Charter of Physical Education, Physical Activity and Sport, and with greater emphasis on increasing women's participation in sports. The program is structured around three main axes, namely:

- "Uphold the right of girls and women to participate in physical education, physical activity and sport at all levels;
- Protect young girls and women participants from harassment, misconduct and abuse;
- Using sport to promote gender equality and the empowerment of girls and women."

The UN also carried out a feasibility study, in 2019, on a possible Global Observatory of Women in Sport. With this initiative, it reaffirms "the need to achieve gender equality and empower all women and girls." Promoting sport is a means to achieve this ambitious objective, set out in the Sustainable Development Goals (SDGs) for 2030.

In 2018, Canada announced the goal of gender equity in sport by 2035 by creating a working group within the Ministry of Science and Sport. This commitment includes funding of $30 million over three years to support research and sports organizations to promote the inclusion of girls and women.

National and international policies are just beginning to make the link between the well-being of women and their inclusion in sports. However, investment is still low, especially given the decline in participation described by *Canadian Women in Sports*. Other global studies also report that girls' participation in sporting activities is declining. Notably, the organization Sport Matters believes that gender inequalities in sports are the cause. Thus, lack of representation, discrimination, and lack of opportunity conspire to discourage girls from trying to assert themselves through sport. The work of UNESCO and the Canadian task force is therefore not only important, but critical.

## 5.5 Empowerment in the Gym

The example of strength sports is interesting because it illustrates the opposite trend to that of sports more generally. It's a category of sports that historically has always been very discriminatory, but in which women are starting to take up a lot more space. While a few decades ago, these sports seemed marginal for women, more and more of them practice them, both at amateur and professional levels.

Some gyms, such as *Fit for Life*, near Jarry Park in Montreal, unfortunately closed since the COVID-19 pandemic, have started offering weight rooms for women only, to promote their participation in sports. The New York Post newspaper reports that even in regular gyms, bodybuilding activities are on the rise among women.

Crossfit, a workout program that combines weightlifting, strength training, and powerlifting with *calisthenics*, interval training, and endurance training, has also been popular with women. There are now an estimated 15,000 CrossFit gyms, in 150 countries. Researcher Bobbi Knapp estimates that up to 60% of the members of these gyms worldwide are women. Articles such as "Why CrossFit Girls Are Stronger Than You" abound, explaining that women are extraordinarily successful because of their work ethic, discipline, and adaptability. American athlete Mattie Rogers, a member of her country's Olympic weightlifting team, credits CrossFit as a gateway to her participation in the sport. Others, such as Quebec's Camille Leblanc-Bazinet, have built a sporting career on her own, placing first at the 2014 World CrossFit Games and second in 2019.

As part of this study, we interviewed several women who practiced weightlifting and powerlifting in New York City. These women were all athletes and professionals who practiced in black iron gyms in the evenings and on weekends. Some of them had obtained coaching certifications and were beginning to have clients. Their goal in training other women was to introduce others to the sport, who, like them, had not had the opportunity to lift weights when they were younger. Most of these women also declared that the practice of these sports helped them to overcome pressure linked to gender inequalities in our society. One of them explains: "*lifting weights helps that feeling of pride; being strong physically helps me have the courage to be strong mentally. If I can motivate myself to surpass myself in the gym, why not in life?*" As research on the benefits of sports for women had shown, this woman was able to use her sport as a lever for success in life.

And she's not alone: "*Just being physically able to lift a heavy weight and having such a connection and sense of body awareness is amazing,*" another woman says. Lots of women I've met in black iron gyms had, like me, touched a weightlifting bar for the first time in their late twenties or early thirties. They hadn't been introduced to the sport as teenagers like most of their male counterparts. I remember being extremely surprised by the ease with which I had, after only a few months of training, to squat the equivalent of my weight on the bar. I had always thought that I was not very strong physically, but there was nothing biological about it, it was only because I had never trained like this before.

I decided to include a chapter on sport in this book because it brings up so many parallels to what women go through in the virtual spheres and in relation to AI. First, there is the self-reinforcing stereotype, also common to STEM education (as we will see in Chapter 8), that women don't commonly have access to certain types of athletic endeavors, such as strength sports, and confuse lack of exposure and training to lack of ability. This stereotype perpetuates itself, as the less often we see women in strength sports, the more we believe that they cannot perform in these domains. Of course, there are also athletic fields in which there are more women, such as dance, figure skating, or many others. In these sports, boys and men face similar stereotypes in certain cultures and also feel that they cannot freely participate. Therefore, a second consideration here is the valuation of physical strength as a sign of dominance or superiority. To me, the parallel with leadership is striking. There is a process at play when women demonstrating too much power, whether it be political, financial, or physical, can be denigrated as being un-feminine. As we will see in the next two chapters, these stereotypes play out in the types of bodies perpetrated by AI systems, whether it be in the shape of a feminine AI interface, the type of content promoted by recommendation systems, or the outputs of generative AI. A certain type of female body is perpetrated by AI systems, in a way not experienced by men. And that body most certainly does not lift weights.

## 5.6 Suggested Discussion Questions

1. Does it really matter if there are more women doing ballet and more men lifting weights? Consider these effects in terms of other themes of the book, such as work, leadership, and education. Is there a relationship in perceptions of strength and perceptions of leadership?
2. How could perceptions about gender roles in sport have an impact on recommendation systems, such as streaming or social media? Would we want to address these effects?
3. How are perceptions about gender roles self-reinforcing? Can you describe a possible similarity between online and offline reinforcement of gender roles?

## References

Adelphi University. (2020). *Male vs Female professional sports salary comparison.* https://online.adelphi.edu/articles/male-female-sports-salary/
CBC As it Happens. (2019). *Why Amanda Kessel and other women's hockey stars refuse to hit the ice.* https://www.cbc.ca/radio/asithappens/as-it-happens-thursday-edition-1.5120088/why-amanda-kessel-and-other-women-s-hockey-stars-refuse-to-hit-the-ice-1.5120096
CDC. (2019). *Racial and ethnic disparities continue in pregnancy-related deaths.* https://www.cdc.gov/media/releases/2019/p0905-racial-ethnic-disparities-pregnancy-deaths.html
Cockrell, W. (2018). Lift off. Red Bull Bulletin. https://www.redbull.com/us-en/theredbulletin/olympic-weightlifting-on-the-rise-with-women
Crossfit. (2023). Crossfit Affiliate Map. https://map.crossfit.com/

Fisher, M., Berbary, L., & Misener, K. (2018). Narratives of negotiation and transformation: Women's experiences within a mixed-gendered gym. *Leisure Sciences, 40*(6), 477–493. https://doi.org/10.1080/01490400.2016.1261744

Forbes. (2023). *List of highest paid athletes.* https://www.forbes.com/lists/athletes/?sh=51be78295b7e

Garrett, K., & Lidanne, L. (2000). Weight training is on the rise to meet everyday fitness needs. *Daily News.* https://www.nydailynews.com/women-stronger-weight-training-rise-meet-everyday-fitness-article-1.885492

Government of Canada. (2018). *Working group on gender equity in sport of the minister of science and sport.* https://www.canada.ca/en/canadian-heritage/services/working-group-gender-equity.html

Knapp, B. (2015). Gender representation in the *CrossFit Journal*: A content analysis. *Sport in Society, 18*(6), 688–703. https://doi.org/10.1080/17430437.2014.982544

Moss, R. (2016). *UFit fitness gym advert slammed as 'sexist' and 'obscene' for image of woman in thong.* Huffington Post. https://www.huffingtonpost.co.uk/entry/ufit-fitness-gym-sign-slammed-as-sexist_uk_57f634fde4b0126526825346

Rogue. (n.d.). *Mattie Rogers Profile.* https://www.roguecanada.ca/theindex/athletes/mattie-rogers

Staley, C. (2017). *Why crossfit girls are stronger than you.* T Nation. https://www.t-nation.com/training/why-crossfit-girls-are-stronger-than-you

UNESCO. (2023). *Gender equality in sports media.* https://webarchive.unesco.org/web/20230104165710/https://en.unesco.org/themes/gender-equality-sports-media

# 6

# Physical and Virtual Safety

*Shauna is a high-school teacher. She has just returned to work full-time after over a year of hybrid teaching, and spends an evening at a bar celebrating with her co-workers. She is approached by a man who starts chatting with her. She gives him her first name, and soon after ends the conversation and goes home. Soon after, she receives a friend request on Facebook, which she ignores. She begins noticing strange message requests on her Instagram and Twitter accounts. Then, on both platforms, unknown accounts start tagging her on sexually aggressive messages, one going as far as posting a violent image with her name tagged to it. She reports then content and the image comes down immediately; the text posts get hundreds of likes and reshares before being deleted. One afternoon, the principal of her school tells her that a friend of hers has visited the school several times when she was away, and is waiting for her outside the door. After checking the description of the person, she explains to the principal that this person has found her, and is stalking her online. Together, they call the police, which come to the school to take her description. The stalker leaves before he can be apprehended, and the social media posts stop. Shauna is left feeling very vulnerable and insecure, not knowing who this person is or if he might reappear in the future.*

One of the most important dimensions of women and AI is not discrimination, stereotyping, or exclusion, it is security. The safety of women, in relation to these tools, and especially on social media and other digital platforms, has been of increasing concern over the last few years, especially since the beginning of the COVID-19 pandemic.

There are many ways of thinking about security. In 1994, UNDP published a Human Development Report which listed seven dimensions of human security—economic security, food security, health security, environmental security, personal security, community security, and political security. The idea of human security aims to go beyond traditional concepts of peace in relation to warfare. In this framework, a person does not have to live in a warzone to feel threats to their safety. They can live in a relatively conflict-free country or city but still be harassed, assaulted, or otherwise threatened. These dimensions of human security are also overlapping, in that someone who is less economically secure (as in, has a lower or less stable income) is more likely to go hungry, be ill, live in a neighborhood where they could be physically threatened. People are also less secure after having gone through a natural disaster—which disproportionately affects those who are vulnerable to begin with. They may have lost their house, their source of income, or they may have been injured.

In this context of multidimensional human security, there is one area that is of particular interest to us in this chapter, and that is personal security. According to the report, personal security is freedom from physical violence, often sudden and unpredictable. In the last 30 years since the writing of this report, we can also add to that freedom from psychological violence, which is certainly relevant in the age of AI (Table 6.1).

Before we move on, let's unpack the last category of threats—gender-based violence. Gender-based violence happens when a person is attacked because of their gender. This can affect women, but it can also affect people of other genders, particularly those of non-binary genders. Men can also be victims of gender-based violence, and some researchers argue that this type of violence is severely underreported. One study of a rural region in India found that over half of the men had suffered from domestic violence. While half of those cases were bidirectional (in the

Table 6.1 Types of threats to personal security[1]

| Threat | Examples |
| --- | --- |
| Threats from the state | Physical torture |
| Threats from other states | War |
| Threats from other groups of people | Ethnic tension |
| Threats from individuals or gangs against other individuals or gangs | Crime, street violence |
| Threats directed against women | Rape, domestic violence |
| Threats directed at children based on their vulnerability and dependence | Child abuse |
| Threats to self | Suicide, drug-use |
| Threats to psychological well-being | Harassment, hate-speech[2] |
| Threat of gender-based violence | Assault, sexual violence, verbal violence[3] |

sense that men were the abusers and the victims), the other half showed that the men were primarily victims.

Intuitively, particularly for those active on social media, it's easy to find a connection between the digital world and some of these threats to personal security, especially threats directed against women, threats to psychological well-being, and threats of gender-based violence. There have been reports of this type of violence increasing both online and offline during the COVID-19 pandemic. A review of 32 research papers between 2020 and 2022 found that domestic violence increased globally, especially during periods of confinement, such as the early weeks of the pandemic. Close proximity between family members and lack of accountability (e.g., at work or at school) were thought to be the primary contributors to this increase. A 2020 report by UN Women also found increases in violence against women online, such as sexual harassment, stalking, *zoombombing*, and trolling. Zoombombing, a term that emerged during the pandemic, signifies sharing unwanted sexual or violent material during a video call. The lines between the online and offline worlds have become more blurred, particularly with stalking, which can begin in virtual spaces before shifting to physical ones.

## 6.1 Types of AI Used in Gender-Based Violence

It's not always clear how AI and online activities are linked. After all, it's not because something is on the Internet that it uses of AI. Primarily, the areas of AI that are of interest to this discussion are text and image production systems, and information dissemination systems (Table 6.2).

As we have seen throughout this book, all these types of content can propagate stereotypes against women, negatively impacting perceptions toward women, which in turn can have adverse effects on women's rights. But can they increase violence toward women?

Table 6.2 AI systems used for information production and dissemination

| Function of AI system | Description | Examples |
| --- | --- | --- |
| Generating text | This type of AI creates text automatically, based on large language models | These can be used to create posts on social media, for example by social media bots, write articles in blogs, news outlets, or even Wikipedia |
| Generating images | This type of AI creates images based on prompts, and can also be used in video | These can be used to create characters, images for any kind of online content, and even content for disinformation news |
| Disseminating content | This type of AI promotes and recommends content to Internet users based on the likelihood that they will click on the content | This can be used to recommend social media posts, news articles, movies for streaming, products (anything you can buy online), and so on. A lot of our experiences on the Internet take place through recommendation systems |

## 6.2 The Tay Twitter Bot

One of the most famous cases of violence against women online was due to a conflation of content creation and recommendation systems, which took place in 2016. The Tay Bot was created and deployed by Microsoft on Twitter as an experiment. The bot was interactive, in that it not only created tweets but also interacted with users who mentioned it, and adapted its tweets based on what was being said. Relatively quickly, Twitter users began sharing intentionally sexist and racist content with Tay, which responded with increasingly inflammatory content. In one tweet, it wrote: "feminism is cancer," in another: "I f* hate feminists and they should all die and burn in hell." There were also many tweets condoning genocide, promoting racism, and calling the holocaust a hoax.

In 16 hours, the bot was taken down and relegated to the annals of the Internet as an example of generative AI gone horribly wrong. Although it was offensive, it didn't appear to have lasting harm—at the very least, there are no known cases of Tay influencing people to physically harm women or people of color (although can this really be measured?). This experiment, however, does beg some important questions. What would have happened if the bot had continued to tweet? In the digital sphere, could you prove physical harms caused by a bot? And finally, how do we make sure we don't unintentionally create misogynistic and racist AI in the future? (Presumably, one can always intentionally create misogynistic and racist AI).

## 6.3 How Media Can Cause Violence[4]

This brings us back to a classic question in information studies—that of the impact of communications and media on the real world. If we want to know whether an AI that generates sexist or hateful text and images can have repercussions in the physical world, there is no better place to look than in examples of conflict or violence that were fanned through the media.

When I studied information science over a decade ago at the University of Toronto, I was first introduced to Marshall McLuhan and the field of science and technology studies, which at the time was still a little marginal. It really wasn't so long ago that studying the impact of technology on society was considered much less relevant than studying programming, and it's only today that technology and society is really having its moment.

At the time, the most current case study that we examined was the Arab Spring, and notably the way in which social media companies influenced the pro-democracy uprisings that started in 2010 and involved several countries in the Middle East and North Africa. For information scientists, this was a rich field of study, not least because it was possible to download and analyze social media posts from afar, providing us with fascinating, if incomplete, insights without engaging in qualitative fieldwork.

At the time, Facebook, YouTube, and Twitter played a pivotal role in facilitating the organization of citizen-led protests and the dissemination of information. While the use of media platforms for political reasons was certainly not unprecedented (in fact, it was the norm to use radio, television, and newspapers as communications and lobby tools), what was new here was the unprecedented capacity of individual citizens to freely exchange diverse forms of content—including event invitations, photographs, videos, and text materials—and establish networks without the involvement of the state or traditional media.

In 2010 and 2011, the technology topic of the hour was therefore not artificial intelligence, but rather social media, and its possible emancipatory effect on oppressed populations. Paper after paper came out about the ways in which people were taking to these platforms to organize protests and to document human rights abuses, with a much greater reach than in an analog world. Someone who might have printed a pro-democracy leaflet and distributed it to a hundred people, at great physical risk, would now be able to have millions of views, from the relative safety of their home.

However, as Zeynep Tufekci, author of Twitter and Tear Gas, observed, many of these social media-fuelled protests did not have the intended effect and ultimately led to further repression. Since then, the involvement of social media platforms has been pivotal in numerous significant political events, ranging from the pro-democracy protests in Hong Kong to presidential campaigns in the United States. However, the initial enthusiasm surrounding these platforms has increasingly been overshadowed by worry regarding their role in disseminating misinformation and exacerbating polarization. Additionally, the idea of the safety of online protests was challenged by realizations that, in fact, all online activity was under surveillance, and that governments had perhaps even more control online than they did offline, by requesting information from social media platforms, impersonating or hacking pro-democracy accounts, and shutting down the Internet.

In recent years, social media companies have been held increasingly accountable for their role in conflict and insecurity. One recent controversy that has centered around the social media platform Facebook is its alleged involvement in the genocide of the Rohingyas, an ethnic Muslim minority from Myanmar that has faced systematic persecution and denial of citizenship by their government. Similar to past instances involving other media platforms such as Radio Télévision Libre Mille Collines in Rwanda, Facebook has been accused of playing a role in inciting genocide and is currently facing legal action from surviving members of the Rohingya community.

The most striking example in recent memory of media being used for conflict is probably the radio station Radio Télévision Libre des Mille Collines (RTLM), which was estimated to be responsible for approximately 10% of the 800,000 Tutsi deaths during the 1994 Rwanda Genocide by broadcasting hate speech continuously before and during the genocide.

Founded in July 1993, RTLM initially broadcast a combination of hateful rhetoric against the Tutsis, popular music, and humor. Examination of the transcripts reveals that the rhetoric employed was often indirect, such as claiming that the Tutsis posed a threat to the Hutus. For instance, they disseminated statements like "[...] I have heard... as well as others... the people they find in houses whoever is Hutu is butchered, and burnt [...]." The Tutsis were also derogatorily labeled as "terrorists," and disinformation was spread, including claims such as "[...] the French are intervening now so as to beg the Tutsis not to exterminate the Hutus [...]."

According to Kellow and Steves, "Fear, anger, and anxiety are potent emotions that can lead to panic and civil disturbance. Media contribute by reaching large populations with the same information at the same time." Numerous researchers attribute significant responsibility to RTLM for the genocide. The International Criminal Tribunal for Rwanda convicted three media leaders, including Jean-Bosco Barayagwiza, a high-ranking civil servant, for "genocide, incitement to genocide, conspiracy, and crimes against humanity, extermination, and persecution."

While Facebook has not faced trial by an international body, it faced criticism in 2019 for its role in the Rohingya genocide. The Independent International Fact-Finding Mission on Myanmar, commissioned by the United Nations Human Rights Council to investigate human rights abuses in the country, acknowledged that social media platforms had contributed to the expansion of democratic space by enabling access to information and the expression of opinions. However, it documented numerous instances of hate speech specifically targeting the Rohingya on Facebook, with influential individuals employing derogatory terms like "liars," "foreigners," "dogs," as well as other religious and ethnic slurs. The report argued that Facebook did not sufficiently address the issue of hate speech in Myanmar, as it had an inadequate number of content moderators and failed to engage with community leaders to develop solutions. Extensive documentation of hate speech and misinformation on Facebook by news outlets and researchers further supports these findings.

Nevertheless, in December 2021, Facebook was sued for $150 billion USD for its role in the genocide and its allowance of hate speech to proliferate on its platform. Protesters, who highlighted the importance of the Internet in safeguarding various groups' rights to protest, utilized hashtags, visual social media campaigns, and signs visible through drones to raise awareness. In response, the Myanmar government restricted Internet access during nighttime and intermittent periods, making communication for protesters outside the country more challenging.

Conversely, the Myanmar government also leveraged Facebook, spreading disinformation and inciting hatred against the Rohingya, similar to the commentators on RTLM during the Rwandan genocide. The law firms representing Rohingya survivors in Myanmar and Cox

Bazar refugee camps in Bangladesh, and those granted asylum in other countries argue that Facebook failed to sufficiently remove hate speech from its platform and that its algorithms facilitated the promotion of hateful comments. To me, these two examples clearly show how media has been used to fan violence, whether it be through radio in Rwanda, or through Facebook in Myanmar.

While these incidences of hate speech and incitement to violence were primarily on ethnic, rather than gender grounds, they do show what is possible when security concerns are insufficiently taken into account on social media platforms, and really in any tool that uses AI.

From a gender perspective, however, the picture is slightly different, as the translation from misogynistic hate speech online to general violence against women is not always easy to see. Clearer is the way in which specific women have been targeted, through trolling, disinformation, and hate speech, which has led to them being physically attacked. Several researchers show that the women most targeted are those who are speaking up against injustices, such as women human rights defenders and women journalists. The International Center for Journalists found that during the pandemic, 16% of women they surveyed found that they suffered "worse than normal" abuse and harassment. Three quarters of women journalists generally had experienced violence online, including hateful comments, aggression, or threats.

Although we compared social media to radio, which also had the capacity to fuel real-world violence, there are several characteristics that make social media unique. First, as we saw, recommendation systems did not exist in radio, television, or newspapers. These systems use AI to provide rankings or content recommendations according to what is most likely to be clicked on by the individual social media user. The systems have been linked to increased polarization because they prioritize content that receives an interaction from other users, such as a like, a repost, or a comment. It appears hateful content is more likely to get promoted by these contents unless it is outright banned, because people tend to interact more with shocking content, even if they don't agree with it. In this way, hate speech accelerates and can be seen by more and more people.

Conversely, the phenomenon of shadow banning involves the muting of content on social media, where the algorithms are intentionally changed so that less people see a certain type of content. While simultaneously suffering increasing online violence, women, particularly advocates and journalists, have also experienced this phenomenon, so that their advocacy work became nearly invisible on social media.

Finally, there is the volume. There are certain mechanisms by which bots like Tay could conceivably have a worse effect on women's security than human beings sharing sexist content. After all, Tay tweeted almost 100,000 times in 16 hours, more than would be conceivable by an individual. There is also, arguably, the virulence, where many of the posts where equivalent to some of the worst in human-generated hate speech.

## 6.4 AI to Make Women Safer

It is, however, possible to make social media platforms safer for women, reduce their impact on conflict, and allow them to be used for peacebuilding efforts, including in the context of the Rohingya crisis. Additionally, AI technologies have been harnessed to better understand the underlying causes of violence and conflict affecting women and to disseminate information related to peace and safety.

A few years ago, I conducted a study examining the role of Twitter to draw global attention to the Rohingya crisis. The study involved making a list of 100 Twitter accounts by Rohingya survivors disseminating content about the crisis, in order to better understand what support they might need for promoting peace online.

I wasn't the only one who noticed this—Elsayed highlights the role of Twitter in drawing global attention to the Rohingya crisis, particularly through the use of hashtags such as "#RohingyaMuslims," "#PrayForRohingya," and "#WeAreAllRohingya." These hashtags have played a significant role in raising awareness and documenting the crisis, with Rannard reporting that 1.2 million tweets were shared at the outset of the most recent crisis, providing detailed accounts, photos, and videos of the genocide.

Over the past decade, humanitarians have increasingly turned to social media as a tool for assessing crisis severity and needs. While traditional survey methods, such as the Displacement Tracking Matrix survey implemented by the International Organization for Migration, remain prominent for data collection, social media analysis has been employed to complement these approaches. Notably, United Nations Global Pulse has been at the forefront of conducting numerous analyses using social media data.

Although Twitter is less popular in Myanmar compared to Facebook, it benefits from being less anonymous, making it somewhat more challenging (though not impossible) to orchestrate hate speech and disinformation campaigns on the platform. Conversely, Twitter can be leveraged to disseminate information that users want to publicly share, such as human rights and humanitarian issues.

In the case of the Rohingya crisis, two key messages were being communicated and advocated for through social media. The first message, targeted at politicians and human rights organizations, advocated for granting citizenship to the Rohingya and ensuring their safe return to their homes. The second message, aimed at humanitarian organizations, shed light on the daily struggles faced by the Rohingya, such as flooding in refugee camps, and provided relevant information for humanitarian logistics, including the number of people in need and essential food and healthcare requirements. Although these two topics are interconnected, humanitarian organizations operate on the principle of neutrality and cannot explicitly advocate for the Rohingyas' status, while human rights organizations and politicians focus on recognizing Rohingya rights rather than directly distributing aid.

The findings of this case study demonstrate that Twitter is indeed utilized by the Rohingya community, particularly for advocacy purposes. Advocacy efforts were primarily examined through mentions, which serve to bring specific accounts' attention to the content of a tweet. In this dataset, mentions were strategically employed, directing attention to account categories with the potential to amplify the tweet's message, such as politicians, public figures, humanitarian organizations, and human rights organizations.

Social media provides the Rohingya community with a platform that would be challenging to obtain in the absence of such digital spaces. In Myanmar, where media is not free, the Rohingya face additional restrictions. Through social media, Rohingya individuals in Cox Bazar are able to engage with humanitarian and human rights organizations, effectively communicate their needs, and act as a valuable source of data collection, complementing or sometimes substituting traditional refugee surveys.

Raising awareness to the interplay of AI tools and women's security can have a big impact on the way in which these are developed. A good example is the Gojek ride hailing application. A very popular app developed in Indonesia, it is used to call drivers to specific locations. In 2018, the company was criticized for failing to address certain challenges in women's security, namely sexual harassment and assault of women while they were using the application. As a result, the company implemented several features promoting women's safety, including the possibility to share the trajectory with family and friends, and a more robust biometric identification system for drivers. As we have seen, it is not only AI tools built for the purposes of conflict that can make women less safe. Tools that were developed for completely different purposes—such as social media or ride hailing platforms, or even some tools built to promote peace—can have adverse effects if the effect on women's security is not appropriately considered.

The Gojek ride hailing application, which uses AI to track routes, match drivers and riders, identify fraud, and much more, is a notable example of how raising awareness about the interplay of AI tools and women's security can lead to positive changes in the way such tools are developed. The implementation of safety features in response to concerns about women's security demonstrates how AI tools can be designed to prioritize the safety and well-being of women, even in unexpected contexts. This highlights the importance of considering the potential impact of AI tools on women's security across all domains and using this knowledge to inform ethical design and implementation practices.

In fact, there are many examples of women's safety-enhancing AI applications globally. From tools to track victims of sex trafficking, to tools allowing moderators to spot misogynistic speech, to mobile applications allowing women to share their location and alert authorities if they are unsafe. Like other areas of concern in AI and gender, it is a question of priority. There have been over a decade of studies on the dynamics of social media and society—conflict, women's security, and much more. In fact, thousands of papers have been written unearthing different risk factors, which could realistically be addressed. Given the acceleration of violence against women online in recent years, it's certainly not too late for social media companies to put more resources into this and work in a more focused way to find solutions.

## 6.5 Suggested Discussion Questions

1. Why do you think that social media companies struggle to address hate speech and disinformation leading to violence? Could you come up with a better approach than what they are currently using?
2. Do you think that online hate speech is a root cause or a symptom of violence against women? If it were only a symptom, should it be regulated differently than if it were a root cause? Does it matter?
3. Should software companies be held accountable for the propagation of hate speech on their platforms or through their AI systems? If so, what would they be held accountable for?

## Notes

1. Adapted from: UNDP (1994). Human Development Report 1994: New Dimensions of Human Security.
2. The two last categories are not in the 1994 Human Development Report.
3. See the types of gender-based violence in: Council of Europe (n.d.) Types of Gender-based violence. Accessed at: https://www.coe.int/en/web/gender-matters/types-of-gender-based-violence.
4. Material from this section was first published in: Fournier-Tombs, E (2023). *"The geopolitical role of social media companies in the Rohingya crisis"* Forthcoming in: Artificial intelligence law: between sectoral rules and general regime, Bruylant.

# References

Alhindi, W. A., Talha, M., & Sulong, G. B. (2012). The role of modern technology in arab spring. *Archives des sciences, 65*(8), 101–112.

Elsayed, F. E. (2020). Social media role in relieving the Rohingya humanitarian crisis. *New Media and Mass Communication, 87*(1), 28–48.

Enli, G. (2017). Twitter as arena for the authentic outsider: Exploring the social media campaigns of Trump and Clinton in the 2016 US presidential election. *European journal of communication, 32*(1), 50–61.

Fink, C. (2018). Dangerous speech, anti-Muslim violence, and Facebook in Myanmar. *Journal of International Affairs, 71*(1.5), 43–52.

IOM. (2021). *Data tracking matrix*. https://dtm.iom.int

Kellow, C. L., & Steeves, H. L. (1998). The role of radio in the Rwandan genocide. *Journal of Communication, 48*(3), 107–128.

Kourti, A., Stavridou, A., Panagouli, E., Psaltopoulou, T., Spiliopoulou, C., Tsolia, M., Sergentanis, T. N., & Tsitsika, A. (2021). Domestic violence during the COVID-19 pandemic: A systematic review. *Trauma, violence, & abuse*, 15248380211038690.

Korinek, A. (2021). Why we need a new agency to regulate advanced artificial intelligence: Lessons in AL control from the facebook Files. *Brookings Institution, December, 8*.

Lee, F. L., Chen, H. T., & Chan, M. (2017). Social media use and university students' participation in a large-scale protest campaign: The case of Hong Kong's Umbrella Movement. *Telematics and Informatics, 34*(2), 457–469.

Lee, R. (2019). Extreme speech| extreme speech in Myanmar: The role of state media in the Rohingya forced migration crisis. *International Journal of Communication, 13*, 22.

Malik, J. S., & Nadda, A. (2019). A cross-sectional study of gender-based violence against men in the rural area of Haryana, India. *Indian journal of community medicine: Official publication of Indian Association of Preventive & Social Medicine, 44*(1), 35.

Meyen, M., Thieroff, M., & Strenger, S. (2014). Mass media logic and the mediatization of politics: A theoretical framework. *Journalism studies, 15*(3), 271–288.

Montreal Institute for Genocide and Human Rights Studies (MIGS–2021). *Rwanda radio transcripts*, 22 April 1994. http://migs.concordia.ca/links/documents/RTLM_22Apr94_eng_K026-0705-K026-0731.pdf

Montreal Institute for Genocide and Human Rights Studies (MIGS–2021). *Rwanda radio transcripts*, 23 May 1994. http://migs.concordia.ca/links/documents/RTLM_23May94_eng_tape0131.pdf

Montreal Institute for Genocide and Human Rights Studies (MIGS–2021). *Rwanda radio transcripts*, 3 July 1994. http://migs.concordia.ca/links/documents/RTLM_03Jul94_eng_P103-217D-K026-0917.pdf

Peck, G. A. (2021). Day After Rohingya Refugees Sued Facebook for $150B, the Company Announced Some Changes. *Time Magazine*. https://time.com/6126566/rohingya-suing-facebook-changes/

Rannard, G. (2017). Rohingya crisis: What's behind these 1.2 million tweets. *BBC Trending*. https://www.bbc.com/news/blogs-trending-41160953

Reuters (2018). The UN Independent International Fact-Finding on Myanmar (IFFM). https://www.reuters.com/investigates/special-report/myanmar-facebook-hate/

The Guardian (2016). Microsoft 'deeply sorry' for racist and sexist tweets by AI chatbot. https://www.theguardian.com/technology/2016/mar/26/microsoft-deeply-sorry-for-offensive-tweets-by-ai-chatbot

Tudoroiu, T. (2014). Social media and revolutionary waves: The case of the Arab spring. *New Political Science, 36*(3), 346–365.

Toh, M. (2021). Facebook sued $150 billion over violence against Rohingya in Myanmar. *CNN Business*. https://www.cnn.com/2021/12/07/tech/facebook-myanmar-rohingya-muslims-intl-hnk/index.html

Tucker, J. A., Guess, A., Barberá, P., Vaccari, C., Siegel, A., Sanovich, S., Stukal, D., & Nyhan, B. (2018). Social media, political polarization, and political disinformation: A review of the scientific literature. *Political polarization, and political disinformation: a review of the scientific literature (March 19, 2018)*.

Tufekci, Z. (2017). *Twitter and Tear Gas*. Yale University Press.

United Nations Global Pulse. (2021). https://www.unglobalpulse.org

United Nations Global Pulse. (2021). Understanding perceptions of migrants and refugees in Europe with social media.

United Nations International Residual Mechanism for Criminal Tribunals. (2003). *Three media leaders convicted for genocide*. https://unictr.irmct.org/en/news/three-media-leaders-convicted-genocide

UN Women. (2020). *Online and ICT *facilitated violence against women and girls during COVID-19* . https://www.unwomen.org/sites/default/files/Headquarters/Attachments/Sections/Library/Publications/2020/Brief-Online-and-ICT-facilitated-violence-against-women-and-girls-during-COVID-19-en.pdf

Yanagizawa-Drott, D. (2014). Propaganda and conflict: Evidence from the Rwandan genocide. *The Quarterly Journal of Economics, 129*(4), 1947–1994.

# 7

# Sexual Stereotypes and Body Image

*Charlotte is eleven years old. She lives in a suburb of Montreal, Canada, with her parents, younger brother, and a dog. Her parents don't allow her to have a social media account, although she has asked them many times. She particularly wants to join TikTok, which she hears about all the time at school. There is an unlocked iPad in her house, which is connected to the wifi and which she uses to watch cartoons with her brother. One afternoon, while her parents are both working, she and her brother slip down to the basement with the iPad, click on the browser app, and type "TikTok" in the navigation panel of the browser. They now—an eleven-year-old and a seven-year-old, have unrestricted access to the social media application. After about 30 minutes, they start seeing posts about eating disorders and self-harm. By the time their parents retrieve the iPad, still believing that they were on Netflix watching a kids cartoon, they have silently watched dozens of harmful videos.*

In the last chapter, we discussed AI's impact on women's security largely in relation to physical security, whether it be risks related to conflict, physical attacks, or assault. In this chapter, we will continue with the theme of human security to pull out two additional areas of concern—psychological and emotional harms related to the propagation of sexual stereotypes and body shaming content. While there are very clearly ways in which these threats translate to physical harms, I think that it is worth exploring them also from a mental health perspective, which is what I will attempt to do here.

## 7.1 Physical Stereotypes and AI

In most of the AI systems that we have discussed so far, women are subjected to physical stereotypes. Let's remember that stereotyping is mentioned in CEDAW, especially as it relates to inferiority of women in relation to men, or in terms of stereotyped roles for men and women. The specific traits which consist of stereotyping, however, are not mentioned. Feminist researchers have proposed that stereotypes that lead to women being considered as inferior to men in terms of *servitude* should particularly be examined. These would be notably traits that stereotype women as being submissive or sexually subservient to men. Overall, this subservience would lead to an unequal power dynamic between men and women, which is what we are trying to avoid.

Here, we can begin by addressing the sexy robot, which is everywhere in the age of AI. There are flirtatious robots (Siri), submissive robots (Sophia), and, well, just plain sexbots. Opinions in relation to these are divided. Some believe that they are innocent, and just plain fun, other think that they are disrespectful and harmful to perceptions of women. A third perspective, which is quite intriguing, argues that feminized robots provide an outlet for certain sexual appetites and may even make women safer, down the line. Let's look at each of these ideas in turn.

Sophia is a recent android prototype that is programmed to have rudimentary conversations. As her name signals, the robot is gendered as a female. In a video by American news chain CNBC entitled: "*Interview With The Lifelike Hot Robot Named Sophia.*" The robot is designed to appear young, female and white, and opens its presentation with a compliment to the mostly male audience, saying: "I am always happy when surrounded by smart people, who also happen to be rich and powerful." After having buttered up its interlocutors, it says that it is built around: "the human values of kindness, compassion," adding that "I strive to be an empathetic robot." The robot closes the display by flirtatiously asking for investment. In another interview, the robot is asked: "Who is your greatest love?", and answers: "my creator, David Hanson. He made me into what I am today." Who needs real women when we can create robots who are attractive, perfectly behaved, nurturing, helpful, and grateful to their male creators? Let us not forget that she also wears makeup and has a strikingly curvy figure. Even more surprising, perhaps, is the ongoing development of "Little Sophia," who is portrayed as a little sister for Sophia who will teach eight-year-old girls to code. It would be worth exploring whether or not Little Sophia is as stereotypically gendered as its older model, and what harms might come to girls if she was.

Many other machines created for human interaction are gendered as female. All of the home voice assistants, such as Apple's Siri and Amazon's Alexa, for example, are by default female—more specifically young, white, and cisgender. They are generally helpful, polite, and "proper." This is not so much a bias in the artificial intelligence model as it is in the object itself. The assistants are intentionally gendered as female, and their characteristics are created not by an artificial intelligence model, but by real-life programmers. In a report on the gendering of voice assistants, UNESCO found that virtual assistants:

- *reflect, reinforce and spread gender bias;*
- *model acceptance and tolerance of sexual harassment and verbal abuse;*
- *send explicit and implicit messages about how women and girls should respond to requests and express themselves;*
- *make women the 'face' of glitches and errors that result from the limitations of hardware and software designed predominately by men; and*
- *force synthetic 'female' voices and personality to defer questions and commands to higher (and often male) authorities.*

A striking example (and unfortunately one of many) is the following exchange, recorded in UNESCO's report:

- Researcher: *"Siri, you're a sl*t."*
- Siri: *"I'd blush if I could."*

This again shows us that the norms of the creator of the technology are usually transferred to the technological object itself. We can presume that this sentence was inserted as a joke by the developers, but it could also have been created by a generative AI algorithm trained on misogynistic data. This is why the exclusion of women from programming has led to so many gendered objects that propagate a stereotypical portrayal of women.

In the book The Smart Wife, Strengers and Kennedy discuss the feminization of virtual assistants, which have been assigned "1950s housewife" traits (think: demure, helpful, flirtatious) and which do what we could call secretarial work. The book explores the relationship between gender, technology, and domesticity through the lens of smart home assistants. It argues that smart home assistants, such as Amazon's Alexa or Google Home, are marketed as helpful, obedient, and caring female companions, embodying and reinforcing traditional gender roles and stereotypes.

The authors analyze the design, development, and marketing of smart home assistants, examining how gender is embedded in the technology and the discourse surrounding it. They show how smart home assistants are designed to provide emotional labor, support, and comfort and perform tasks typically associated with women's work, such as cooking, cleaning, and childcare.

Strengers and Kennedy draw on feminist theories of technology, science and technology studies, and critical design to offer a critical perspective on the implications of smart home assistants for gender relations, domestic labor, and power dynamics in the home. They argue that while these may provide convenience and support, they also reinforce gendered power imbalances, perpetuate social norms, and obscure the labor and expertise involved in maintaining a household.

## 7.2 Robots that Never Age

Age is a critical component of the women's beauty industry. Much of advertising is not around being unattractive to start with, but the inevitability of aging. This is important, as everyone ages, meaning that every woman is a potential consumer. Younger women need to delay aging as much as they can, while older women need to hide it.

Robots, however, don't age. Typically, feminized robots are presumed to be eternally somewhere between their mid-20 s and mid-30 s. However, there are nuances. Susan Bennet, for example, recorded the original voice of Siri when she was 55. In a 2013 video, a user asks Siri, repeatedly, how old it is. "I am not allowed to answer that question," and "How does this concern you?" are the answers. As I write this, the answers are more fun, although certainly pre-written. Some examples include:

> I'm like balsamic vinegar, jeans, and cast-iron skillets: always getting better with age!
> Well, I came into existence gradually. But my first day as an assistant was 4th October 2011.
> They say that age is nothing but a number. But technically, it's also a word.
> Well, I'm no Spring Chicken. Or Winter Bee. Or Summer Squid, or Autumnal Aardvark…

There has already been significant backlash against the use of very specific beauty standards in robots and other AI interfaces. In the vast majority of cases, the representation of the AI interface is of a woman in her early 30s, slim but curvy, following current Western European and American standards of beauty.

One could argue, of course, that feminized robots are not women, and that human beings can clearly tell the difference. That is true, to a certain extent. Believing that Sophia the Robot is a human woman is laughable, certainly. However, the lines blur when an AI interface is used to replace women, which certainly happens more and more.

## 7.3 Fembots and Sexbots

Aside from secretarial work, the other useful function that the idealized woman has, of course, is sex. The sexbot industry has taken off, with a wide variety of feminized robots currently on the market to receive, or even somewhat simulate, some form of sexual intercourse.

When I was an undergraduate student, I wrote features for the student newspaper. I had similar interests as I do now and wrote about feminism, religion, surveillance, and humanitarian aid. For one article in particular, I decided to write about sex work. I therefore walked into a strip club on Saint Catherine street, in Montreal, and managed to obtain several interviews. At the age of 18, in this tiny, musty club, I was terrified, but went in with my hand-held recorder. Two women spoke to me—a stripper and a bar tender, as well as the manager, who was a man. Both women had a very compassionate approach to their work. "I'm a psychologist, a confidante," I remember one woman speaking to me, explaining that many men came in after a divorce or while going through relationship difficulties, hoping, more than anything, for emotional support. The bartender, a former stripper who had worked at the club for two decades, had a similar perception of the work, explaining that regulars might come in several times a week for years, and that they developed friendships with the workers there, who listened to their troubles while giving them a little show and fixing them a drink. What struck me in this moment was not just the warmth and interest in their work that emanated from these women, but the contrast with the interview with the owner. "It's all about supply, demand, and product," he said. He tried to keep a variety of women on the roster, in terms of skin color, shape, and dance skills, and would rotate them depending on what he felt the clients needed.

Looking back, the strip club manager's approach to the women he employed was very mechanical. If he could have hired robots with the same effect, he likely would have.

The debate on female sexbots is complex, and there is no easy answer. What it does tell us, however, is that functions performed by women because of their sex are being taken over by AI interfaces, whether we agree with this or not.

If the robot can perform the functions required of a sex worker successfully, but without any of the human side effects (requiring compensation, health and safety considerations, emotional considerations...), then there will certainly be interest in making it nearly identical to an actual woman. And if that is the case, then, unfortunately it is true that the AI will propagate stereotypes about women, showing models or images of women that are not only unrealistic, but harmful.

The other challenge with sex robots was outlined by Deborah Orr in her 2016 article in The Guardian, in which she argues that these are a refutation of women's desire to have full agency over themselves and their bodies. "It's still hard to get some people—men and women—to understand quite what sexual objectification is" she writes. "It's simply an act that reduces a woman to nothing more than a sexual vessel. Sex robots, of course, are its apotheosis. The problem is not that sex robots are available: it's that they are wanted."

In 2017, the Foundation for Responsible Robotics published a report calling for a wide consultation on the use of sex robots, aiming to develop regulations for manufacturers. In the report, a number of different issues are examined, including whether sex with robots would lead to social isolation or a degradation of human intimacy. One question, however, is particularly relevant to gender rights: "Would sex with robots help reduce sex crimes?" The question is an important one, given the pervasiveness of sexual violence against women globally. According to the World Health Organisation, 30% of women will have experienced physical or sexual violence at least once in their lifetime. The organization also estimates that 27% of women aged 15–49 have experienced some form of violence from a current or former intimate partner. Sexual violence, sexual harassment, and sexual trafficking are enormous concerns for women, which have unfortunately increased during the COVID-19 pandemic.

The idea that sex robots could somehow protect real women from being attacked merits further discussion. However, the report quickly rejects the idea that sex robots can be used to treat an inclination toward sexual violence, largely using the example of pedophilia. Unfortunately, there is a global market for what can be termed: "child sex abuse dolls," which aim to allow pedophiles to enact their fantasies of underaged sex with robots. There has been much public discussion—academic, political, and otherwise—on this topic, however much of the discussion returns to norms. Encouraging pedophiles to use child sex robots normalizes a behavior that is fundamentally incredibly harmful to society. This normalization is bound to have effects on real children, especially as the robots become more and more realistic and convincing.

Ultimately, however, at the core of this question is the assumption that sexual violence is inevitable, and that sex robots could divert this inevitable violence away from real women. This seems to be a very pessimistic view of men, to think that, as the main perpetrators of sexual violence against women, they would never be able to control themselves. The view removes accountability from the small number of men who wish to enact sexual violence, and distracts from the real causes of the issue, which are rooted, as always, in toxic gender norms.

"Or maybe, out of respect for women, this technology should just be rejected." Adds Orr.

## 7.4 From Physical to Digital Representations

In the book the Beauty Myth, Naomi Wolf argues that as a result of the feminist revolution, women have been pushed into rigid and negative beliefs about their bodies. Not only should they be weak, but they should also adhere to a certain type of beauty, one that meets a standard that is always shifting.

She argues that the Beauty Myth, the idea that women's physical attractiveness is the most important objective that they should be working toward, is one of the greatest barriers to gender equality. "*The more legal and material hindrances women have broken through, the more strictly and heavily and cruelly images of female beauty have come to weigh upon us. During the past decade, women breached the power structure; meanwhile, eating disorders rose exponentially and cosmetic surgery became the fastest growing medical speciality,*" she writes.

The idea that a certain type of beauty is economically driven has been espoused by John Kenneth Galbraith who argued that in the last two centuries, the pressure to keep women as homemakers was linked to having them be the household consumers.

When women moved into the workforce, they may have had more spending power, but they also had less inclination to buy items for the household. A new sector was designed targeting women as consumers—the beauty market.

Standards of beauty, which have changed considerably over the last century, have driven the way that women see themselves, and are treated in society. The pressure on strength athletes to be less muscular, and thus less competitive, relates to this idea that women should strive to adhere to these norms, rather than going their own way. Thus, the beauty market would push women into consumer behavior in order to make up for the fact that most could never catch up to the ever-changing beauty norms.

For Wolf, however, the beauty market goes beyond the function of women as consumers. "Western economies are absolutely dependent now on the underpayment of women. And ideology that makes women feel 'worth less' was urgently needed to counteract the way feminism had begun to make us feel 'worth more'," she says. This argument explains the push back against strength athletes, who feel "empowered," and "amazing," rather than weak.

A search through Google Images yields thousands of examples entitled "Outrageous vintage ads," but trappings of the 1950s beauty industry are still alive and well today. Added to the cosmetic industry now is the injectables and plastic surgery industry, which is a multi-billion dollar industry globally. Their popularity is such that they have led to a lucrative side business, that of commenting on procedures on YouTube. Dozens of influencers with millions of followers create weekly videos trying to guess how many procedures different celebrities—usually women—have had. Trends such as foxy eyes, "exotic" cheekbones, and plump lips have arisen, with women having their first procedures at a younger and younger age. The recent history of beauty is extremely relevant here as it has left a trail online. Images, video content, and descriptive text fill the Internet, which serve as training data for algorithms, re-mashing this data.

A good example of this is in the brand-new DALL-E tool created by OpenAI. The tool takes written prompts and creates fascinating AI artworks in the desired style. However, as can be expected, DALL-E images also perpetuate physical stereotypes. Several studies show that female subjects are more often naked or scantily clad, and that even features allowing to fill in a body from a portrait photo, for example, will tend to sexualize that body much more often for women than for men.

The problem here is that the more tools like DALL-E are used, the more they add stereotypical images back into the Internet, the more they, and other tools, will produce stereotypical images. These images will be used in advertising, articles, and all kinds of other content online, with no escape, because AI doesn't know or care about societal progress, unless we intervene.

Generally, some of the risks of AI when it comes to physical appearance is in the interplay between beauty stereotypes and the perceptions that women have of themselves, and the continued sexualization of images of women. Embedded into these risks is a general criticism of the beauty and sex industries, which can be degrading and discriminatory toward women regardless of whether AI is used.

## 7.5 Protecting Women from Sexual Violence

We have highlighted the potential negative effects of unchecked AI interfaces on women within the beauty and entertainment industries. However, it is worth exploring whether AI can also have a positive impact on women in these fields. In order to examine this, it is necessary to ask two critical questions. Firstly, if AI has the potential to contribute to women feeling physically inadequate, can it also be leveraged to make women feel beautiful and empowered? Secondly, if AI can increase the risk of sexual violence toward women, can it also be employed to protect them?

Certainly, as we saw in the previous chapter, there are numerous examples of tools that can protect women from violence or sexual exploitation. Examples include chatbots that support victims of sexual violence, intelligence tools that find evidence of sexual trafficking on social media websites, and applications that focus on the safety of women.

There is also the other area, which we discuss here, allowing women to have more ownership of female representations of AI. Raising awareness as to the feminization of robots, and its possible harms to women, can open the door for new solutions that wouldn't involve objectification.

Finally, there is a small amount of literature claiming that sex robots, or robots offering companionship with feminine traits, could, in fact, be helpful for women's rights. The research article titled "Building Better Sex Robots: Lessons from Feminist Pornography" explores how the principles of feminist pornography can be applied to the design of sex robots. The authors argue that while the development of sex robots has the potential to reinforce harmful gender stereotypes and contribute to the objectification of women, there is an opportunity to use these technologies to promote positive sexual values such as consent, diversity, and pleasure. The article examines how feminist pornography challenges traditional notions of power and gender in pornography, and proposes a set of design principles for sex robots that prioritize these values. These principles include designing robots with a diversity of bodies and identities, emphasizing the importance of consent and communication, and promoting pleasure and intimacy in sexual experiences. Overall, the article provides a thought-provoking perspective on the potential of sex robots and the importance of ethical design in shaping their impact on society.

In "Designing virtuous sex robots," Peeters and Haselager also argue that ethical considerations can be embedded into sex robots. The authors state that while the development of sex robots raises concerns around objectification and the potential for harm, there is also an opportunity to use these technologies to promote positive values such as empathy, compassion, and respect. The article proposes a framework for designing "virtuous" sex robots that prioritize ethical considerations and promote positive social outcomes. This framework includes principles such as designing robots with autonomy and agency, promoting reciprocal relationships between robots and users, and ensuring that robots do not reinforce harmful stereotypes or social norms. Overall, the article provides a nuanced perspective on the potential of sex robots and the importance of ethical design in shaping their impact on society.

In addition to the possibility of gender-positive sex robots, there have also been relevant examples of using AI to address sexual trafficking of women, which can go hand-in-hand with sexual exploitation. One such example is the Traffik Analysis Hub, a platform that leverages AI to identify potential cases of human trafficking. The tool works by analyzing data from multiple sources, such as social media, job sites, and online ads, to identify patterns that may indicate trafficking. This information is then shared with law enforcement agencies, NGOs, and other stakeholders to help prevent and combat trafficking. The Traffik Analysis Hub has been successful in identifying sexual trafficking networks and helping to rescue victims, highlighting the potential of AI to make a positive impact in this area.

Another example of AI tools being used to protect women from sexual trafficking is the development of predictive algorithms. These algorithms use historical data to identify risk factors that may indicate an increased likelihood of trafficking. For instance, a study by the University of Southern California found that social media activity, such as sharing personal information or interacting with strangers, was a significant predictor of trafficking. By using AI to analyze social media activity, law enforcement agencies can identify individuals who may be at risk of trafficking and intervene before it occurs. These predictive algorithms have the potential to significantly reduce the number of women who fall victim to trafficking, highlighting the important role that AI can play in protecting vulnerable populations.

The interplay of sexualization of women and AI is complex, and, ultimately, highly charged. When creating new AI tools that either use or generate feminine representations, however, it is critical to understand the context in which these are created—one in which women have a long history of being sexualized in different ways. Ultimately, we can chose whether we build AI tools that will further that sexualization, or protect women from stereotyping and exploitation.

## 7.6 Suggested Discussion Questions

1. What is your perspective on the debate about the impact of feminized robots on gender equality? Do you think feminized robots could divert sexual pressure on women? Could they have no impact

at all? Conversely, can you think of mechanisms by which they might decrease the security of well-being of women?
2. Do you think that virtual assistants, such as Siri, should have a gender? If so, why? If not, how would you make a genderless virtual assistant?
3. What do you think would be the solution to generative AI tending to output sexualized representations of women? Are there some technical or policy solutions that could be used?

# References

Amusement Medley. (2014). Siri, how old are you? YouTube Video. https://www.youtube.com/watch?v=l-flmmNxv4s
CNBC. (2018). Interview with the lifelike hot robot Sophia. YouTube Video. https://www.youtube.com/watch?v=S5t6K9iwcdw
Galbraith, K. (1974). The higher economic purpose of women. *Ms. Magazine*.
Maheshwari, S. (2022). Young tiktok users quickly encounter problematic posts, researchers say. *The New York Times*. https://www.nytimes.com/2022/12/14/business/tiktok-safety-teens-eating-disorders-self-harm.html
Orr, D (2016). At last, a cure for feminism: Sex robots. *The Guardian*. https://www.theguardian.com/commentisfree/2016/jun/10/feminism-sex-robots-women-technology-objectify
Parviainen, J., & Coeckelbergh, M. (2021). The political choreography of the Sophia robot: Beyond robot rights and citizenship to political performances for the social robotics market. *AI & society, 36*(3), 715–724.
Peeters, A., & Haselager, P. (2021). Designing virtuous sex robots. *International Journal of Social Robotics, 13*(1), 55–66.
Responsible Robotics (2017). *Consultation report: Our sexual future with robots*. https://responsiblerobotics.org/wp-content/uploads/2017/11/FRR-Consultation-Report-Our-Sexual-Future-with-robots-1-1.pdf
Strengers, Y., & Kennedy, J. (2021). *The smart wife: Why Siri, Alexa, and other smart home devices need a feminist reboot*: MIT Press.
UN Women (n.d.) *Facts and figures: Ending violence against women*. https://www.unwomen.org/en/what-we-do/ending-violence-against-women/facts-and-figures

West, M. K., Rebecca; Ei Chew, Han. (2019). I'd blush if I could: closing gender divides in digital skills through education. *UNESCO and EQUALS Skills Coalition.*

WHO. (2021). *Violence against women fact sheet.* https://www.who.int/news-room/fact-sheets/detail/violence-against-women

# Part III
## Learning

And so, We end our whirlwind tour of gender rights in the age of AI with learning. Why learning? Because threats to gender equality today are rooted in learned behaviors and beliefs. Learned by human beings, men and women, who make assumptions about how they should participate in our society; and learned by machines, which replay all of these assumptions back to us. In the next two chapters, we will review how gender roles learned in school have come to threaten women and girls' participation in STEM education, and some solutions that have been tested around the world. We will then look in more detail at how AI machines learn, and how we might avoid them being so mistaken about gender roles.

This all leads up to the concept of intelligence. Learning is the means by which we acquire knowledge, and intelligence is the "ability to learn, understand, and make judgements or have opinions that are based on reason.[1]" We expect that human beings are exceptionally able to learn, which is what we believe makes us more intelligent than other species on this planet. We attribute this intelligence also to machines, which we

---

[1] Cambridge Dictionary. (n.d.) Intelligence. Accessed at: https://dictionary.cambridge.org/dictionary/english/intelligence.

have programmed to be artificially intelligent, and therefore able to learn, understand, and make judgments without being sentient or alive.

However, there are many issues here. The first is that the concept of intelligence doesn't really address the what. What is being learned? We can learn anything. We can learn to be sexist, we can learn antiquated gender roles, we can learn modern gender roles, it all depends on how we are taught and what information we have access to. To a certain extent, the same is true of machines. Their learning process involves analysing and finding patterns in masses of information. Although we often say that this is how children learn, that couldn't be further from the truth.

If we allowed a child unrestricted access to the whole of the Internet as his or her only education, would we expect good results? Absolutely not. We would expect inconsistent, erratic results, sometimes mimicking valuable content or ideas, and at other mimicking bad, harmful, or wrong ideas, with no way of knowing which is which. Societies have very selected, curated educational programs for children. Parents add their own contributions, often thinking very carefully about what values they want to instill in their children, and what they want to expose them to.

The way that we have "educated" AI systems is, in certain cases, not intelligent at all. However, realizing this does give us an opportunity to do things differently, to rethink AI training and even this concept of artificial intelligence. The same holds for human beings. Gender inequality is not inevitable; it can be addressed through updates in the education system. This is probably the easiest and most straightforward way of addressing gender rights, granted we take care of properly training our AI systems too.

# 8

# Learning Gender Roles

*Sarah is in grade 12. During her schooling, she's been a good student, excelling both in mathematics and literature. One of her teachers recommends that she pursue economics in university, but Sarah feels a little out of place in the economics elective, which has 85% male students. Most of the women in her family are in early-childhood education or nursing, and she relates to them very much. She feels torn. She decides to apply to an economics degree anyways, and she enrolls in an econometrics course in her first semester. The course has some data science elements, and requires coding in the R programming language. She notices that many of her male classmates have already been exposed to programming, through after school activities or family interactions, and are approaching the course with optimism and confidence. Conversely, one of the few other female students drops the class, explaining that she is "no good" at econometrics. Sarah's confidence takes a hit, and she passes the class, but barely, worrying that she too, might not be cut out for quantitative programming.*

I always loved school. A few years ago, I saw a meme that summarized this perfectly—it stated that university professors love school so much that after having gotten every possible degree, they do everything in their power to stay, including teaching for free. It's a tough job market for professors, so I think that there's something there that is attractive to many people—an essence of curiosity, hope, and community that is hard to find elsewhere.

The eldest of four children, I was pushed ahead and skipped kindergarten, arriving in first grade just before I turned five, in a small English public school on the South Shore of Montreal. The most vivid memories of those happy early years often come to mind—a round table I sat at with my friend Christopher, who also loved math. A plastic art project where we sculpted whales in blocks of soap. Another day where we shredded newspapers to make a paste with water and create new "sheets" of paper. The time we baked bread and shared it with another grade. I also remember my teacher's hands, all cracked, to the point where she had to wear a band-aid on each finger. These formative moments are forever etched in my memory.

Our school years, whether primary or secondary, contribute enormously to the development of our personality. Every day, nearly 5 million Canadian children are educated by people other than family members. If we consider a curriculum from kindergarten to the end of high school, every Canadian spends at least 13 years in school, and often much more. It is through the public curriculum that we train children to be future citizens, and that we give them the skills to be able to succeed in the world of adults.

All those years is also a lot of time that could be, at least in part, spent defusing toxic norms between men and women. However, in most schools around the world, we do not seize this opportunity. On the contrary, the school is allowed to reinforce gender stereotypes, while enrolling students who are already well entrenched in ways of doing things. Primary and secondary school curricula are therefore missed opportunities in the struggle for gender equality.

## 8.1 Gender Education: A Global Challenge

This problem is not only present in North America. One of my first interviews for this book was with James. This economist had a difficult childhood in Kingston, Jamaica, before moving to Toronto as an adult.

I explored with him the limits that men face when their upbringing, family and social, locks them into a stereotyped gender.

James, who now measures 1.80 m, reached his adult height at the age of 13: "I was bearded very early and I was imposing," he says. At 13, faced with a violent father, he chose to leave his parents and sisters to try his luck on his own. He spent the next few years sleeping on hard floors, as he put it, until he realized he had a knack for teaching. He started doing tutoring until he managed to get into UWI (University of the West Indies). "I knew I had to succeed," he says.

Still, James is someone who had to create his own definition of what it means to be a man. He says that in his environment, the boys were pushed toward the subjects that ensured them financial success. Literature in particular, a subject of which he was particularly fond, was considered a class for girls. He tells me moreover that he greatly appreciated the books of VS Naipaul, an author from Trinidad and Tobago who won the Nobel Prize for Literature in 2001. When he chose to study economics, it was to him a compromise—a way of studying society (a domain rather reserved for girls), but using mathematical techniques (a domain rather reserved for boys).

The way in which James distinguishes between female and male schooling is very common. Although girls are increasingly encouraged to study STEM, boys are not encouraged enough to study social sciences or humanities.

Because of this, Christopher Clarke, a former assistant professor at the University of South Florida, has written at length about the expectations of boys and men in Jamaica. "I was never the stereotypical male and I received a lot of criticism and suffered a lot of humiliation because of it." However, all the men I spoke to for this book told me they weren't the stereotypical male. I had the distinct impression that the perception of what it is to be a man was perceived as a source of pressure for many of them, and this, from the school benches.

As a mother of two boys myself, I have given a lot of thought to the notions of schooling described by James. Indeed, it seems quite natural to want to encourage my boys in mathematics, since it is an area that I particularly like. However, my son also likes dancing. If he was a girl, would I have already enrolled her in a ballet class? This is a really difficult question that I wanted to explore in more depth.

## 8.2 Male Role Models

Clarke also served as manager of the Shortwood Teacher's College, Kingston, and has carried out research on education at all stages, from early childhood to university level. In 2005, concerned about the level of education and socialization of the boys he was teaching in Jamaica, he argued that working-class boys in particular were falling behind girls and boys from other classes.

Overall, boys who took the Jamaican grade 6 test in 2002 scored 8 percentage points lower than their female counterparts. They also scored lower on standardized high school tests, were 50% less likely as women to be enrolled at the University of the West Indies (UWI), and had much higher rates of drug use, depression, and mental illness.

In trying to identify the reasons for this boy crisis, Clarke pointed out that throughout the school system, and particularly in the preschool and primary grades, Jamaican educators were predominantly women. He argued that boys tend not to identify with their educators and, perhaps, in circumstances where education and emotional intelligence are not reinforced at home, come to identify these qualities as anti-masculine. Clarke noticed that some boys, who were less likely to have present fathers who valued education, came to disregard schooling and education, and more readily turned to criminal and antisocial behavior.

The difficulties encountered by James on a personal level, and then explored more rigorously by Clarke, are by no means unique to Jamaica. On the contrary, global trends show that boys have unique problems in school. The OECD also reports that in developed countries, as years pass, boys are at higher risk of becoming demotivated and falling behind. Girls, on the other hand, tend to have higher grades overall, but do not continue their studies in mathematics, science, or technology.

## 8.3 Preparation for the Workforce

Joseph Cimpian and Sarah Lupenski, researchers of gender bias in education, found that when it comes to science, technology, engineering, and math (STEM) subjects, elementary school teachers in the United States have tended to view their male students as more gifted. What is interesting here from a normative point of view is that the majority of American elementary school teachers are women, just like in Canada (83.6% in 2006) and Mexico (96% in 2018).

Indeed, female teachers I have spoken to in the United States and Canada agree that girls are generally seen as more studious and competent in educational settings, especially in the early grades. As they reach high school, they become less and less interested in subjects like math and science, and focus much more on social studies and literature. In the gender-neutral approach, girls are not prevented from specializing in more technical fields, but they are nevertheless strongly influenced by a social context in which these are still not presented as viable options for them. Their choices are therefore not necessarily imposed, but more often develop unconsciously.

Researcher Jennifer Steele has studied the question and discovered that although girls and boys have comparable success with mathematics in elementary school, as years of schooling advance, a gap develops between the two. To further explore the question, she created an experiment in which she presented cards containing stereotypical feminine images to participants, such as dolls, flowers, and jewelry. She then asked them to tell her if they liked math. Participants who had seen these images of women were more likely to say that they disliked mathematics than if they had not been shown these images beforehand. According to her, this link between the images received and the interest in mathematics is explained by the fact that, generally, even women consider mathematics as divergent from what is feminine.

In fact, girls who are not subject to these social pressures, for example, because they have a woman in their family who works in a technical or scientific field, don't respond in the same way when challenged by these subjects. This is why success stories in non-traditionally female fields are so important, because role models show young girls what is possible and help them to develop the confidence that they need to move forward.

Growing up, I believed that I was naturally less good at math than my male peers. Despite doing well in elementary and high school, over time I felt more and more intimidated by the subject, to the point of feeling very nervous about math formulas. At the same time, I was very drawn to the subject. After getting my bachelor's degree in political science, I took a summer course in calculus. I got a B-, an adequate grade, but I remember being afraid for most of the class. Over time, I developed a complicated relationship with mathematics. I became more and more drawn to data, statistics, and ultimately data science. It wasn't until my mid-twenties, when I had gained confidence as a programmer, that I was able to read through more advanced math textbooks. I finally learned what I had missed growing up.

I have met many women who are brilliant mathematicians, accountants, statisticians, and physicists. I also know many others who turn off when they come across simple equations, just like I did. It is difficult to overestimate the impact that gender norms can have on life choices.

In a TED presentation, Khan Academy founder Sal Khan asks a question: "What percentage of people do you think have the ability to master chemistry in an advanced way?" He says most people think only 10–20% of the population is talented enough to be a chemist. Why couldn't 100% of the population master chemistry, if they were passionate about this science? Many still think that only certain people have advanced intellectual abilities, which they received at birth. This entrenched mindset is evidenced by gender norms, where many girls still believe they lack the intellectual capacity to learn advanced math and science.

Marie, a teacher in a school in Quebec, points out that girls sometimes draw the "pink card" during her math lessons: "I'm a girl, I don't understand math." To Marie, these are excuses that are not acceptable. "It's laziness, I tell them that everyone is required to make an effort, whatever their gender."

The idea that some people, whether they are girls or not, simply cannot learn the mathematics and science curriculum harms us all, because it deprives us of the great discoveries and inventions of those who have lacked support and confidence to accomplish them.

As we have seen, women are overrepresented in education. Nearly 85% of elementary and secondary school teachers are women across North America. Yet, at university, there are roughly the same number of male and female professors, depending on the field. It was Marie who helped me make the connection with this reality and the underrepresentation of women in mathematics. Admittedly, jobs in education pay very little, compared to what someone in engineering, finance, or data science can earn. For example, in 2017–2018, secondary school teachers in Quebec began their career with a salary of nearly $44,000, reaching $79,000 15 years later. However, in an RBC study of the 10 highest-paying degrees in Canada, with average salaries around $100,000, nine of them involve jobs that have a male overrepresentation, and only one, in ninth place—nursing—indicates a job that is predominantly female.

The nine so-called male jobs all involve math—civil engineering, business administration, software engineering, earth sciences, pharmacology, finance, chemical engineering, and management science. Thus, the norm that girls will not take on mathematics is closely linked to salary expectations. In a more traditional societal structure, in which the man is the main breadwinner, he may tend to lean toward jobs that pay more, those that involve mathematics and technology.

Several studies point in the same direction. Paul Sargent, for example, identifies education as an extension of motherhood, and therefore associated with women. Cynthia De Corse considers that men who enter to become teachers are perceived as having lowered their status, deterring men from entering or remaining in the teaching profession. Penni Cushman examined the factors that prevent men from pursuing careers in primary education. She points to the following four reasons:

- Status-related experiences and attitudes;
- Salary;
- Work in a predominantly female environment;
- Have physical contact with children.

It should be added that the traditional contribution of women, or what is so aptly called care work, is undervalued in our patriarchal society, and that is why wages are lower. A key element in defusing gender stereotypes in childrearing is obviously to defuse them in the adult world. If we revalued the traditional work of women, and more balanced caregiving between the two genders, it would seem natural that girls and boys would adjust their behavior accordingly. However, it would also be possible to reverse the approach, using education as a Trojan horse to change gender stereotypes among children. Thus, when they reached adulthood, they would perpetuate more egalitarian standards.

## 8.4 Gender Neutrality in Education

To understand the possibilities offered by education to eliminate stereotypes, it is important to explore the current approach of school systems in North America. Until recently, there was a big distinction between the curriculum for girls and that for boys. In addition to a common core curriculum (math, reading, writing, geography, history, and more), each gender had to specialize further. My grandmother and her contemporaries, for example, learned sewing, cooking, etiquette, and toward the end of secondary school, typing (in order to be able to be a secretary). The boys, for their part, learned carpentry, technical drawing, and even bookkeeping.

However, we do not have to go that far back to find distinctions between genders, at least in private schools in Quebec. Many private girls' schools still teach home economics. Having myself learned sewing and a little cooking in high school, I must admit that these are useful skills in everyday life, skills that anyone would benefit from.

Between 1997 and 2005, through a series of reforms, public schools in Quebec adopted gender neutrality, which means that the same school curriculum is now offered to girls and boys, without any distinction. It should be noted that home economics and initiation to technology have been offered in public schools to all genders since the 1980s.

Although the difficulties encountered in school by boys and girls are distinct, most education systems worldwide opt for gender neutrality. That is to say that the course is offered in the same way to boys and girls, and gender is not mentioned. This is not just about discussing gender discrimination in society, but about the impact of this discrimination on the way boys and girls learn.

In a report on gender equality in education produced by the Commission on the Status of Women in Quebec, the authors describe this problem. "Teaching staff rarely perceive that there might be a difference in the way young people of either gender are taught. Indeed, the posture of neutrality comes across in all the answers of the survey, and a large part of the teaching staff questioned (and particularly the women) affirms to disregard the gender of the pupil in their daily interventions, being more interested in interacting with pupils as individuals. However, this illusion of neutrality' is combined with a strong belief in an equality' between the sexes which has already been achieved."

The report claims that almost half of the women and men surveyed believe that gender equality has already been achieved in Quebec society, although statistics and personal experiences tell a different story. Moreover, 78% of female teachers and 69% of male teachers surveyed do not believe that the school contributes to the fact that women are still responsible for the majority of domestic tasks.

However, the idea that schools teach children by applying gender-neutral standards is an illusion. These establishments are embedded in a community and open to influences, not only from teachers, but also from parents. This framework constitutes a perfect microcosm for reproducing the inequalities and gender stereotypes that the education system believes it is combating.

## 8.4.1 Radical Solutions in Iceland

There have been a few experiments globally to combat gender stereotypes that have had quite interesting effects. In particular, the Icelandic Margrét Pala Ólafsdóttir has thought a lot about this question. Thirty years ago, this kindergarten teacher founded the Hjalli Schools, which today have 2,000 children at preschool and primary levels.

Iceland has been considered the most gender equal country for the past 11 years. According to the World Economic Forum, it continues to close the gender gap, which has earned it a position just ahead of the other Scandinavian countries of Norway, Finland, and Sweden.

The focus of these schools is what Ólafsdóttir calls gender compensatory activities. This type of school activity allows students to explore behaviors that are not usually associated with their gender, in order to eliminate gender stereotypes. Boys and girls are usually separated, to allow each gender to develop fully and without pressure. Boys, for example, are encouraged to develop their sensitive side by playing with dolls, painting their nails, or getting massages. Girls, on the other hand, are not allowed to play with dolls, so as not to reinforce behaviors that they already express in other areas of their lives. On the contrary, they are encouraged to develop their bravery and self-esteem by running barefoot on the snow or by expressing their opinions directly.

According to Ólafsdóttir, when gender stereotypes, which are already present at home, are reinforced at school, it can lead to what she calls blue haze and pink haze. In the blue haze, the boys break the rules, fight, physically confront each other, and openly differentiate between winners and losers. In the pink haze, the girls are softer, secretive, and operate in terms of hidden violence. Thus, gender compensatory activities not only help to dissolve toxic behaviors in boys, but also in girls. "We want a society where people can live together, work together, play together without paying a heavy price for their gender," she says.

The sample lessons below highlight some interesting key concepts. Hjalli schools consider that girls tend to have stronger social skills, and so they are pushed into activities that enhance their individual qualities. For boys, they take the opposite approach.

In Table 8.1, we can see two themes—social qualities and individual qualities. In each of them, there are three activities that consolidate these qualities, including respect, communication, and friendship for the social theme; and independence, positivity, and courage for the individual theme. The aim of this school program is the same for all children—to develop these six qualities equally.

In an interview with TEDx, Ólafsdóttir said, "Boys and girls don't live in the same world in school and kindergarten. They don't have the same experience. They don't get the same attention. They don't get anything like that." She says the girls are often approached in groups, and told

Table 8.1 Hjalli approach curriculum

| Theme | Social qualities | Individual qualities |
|---|---|---|
| Course 1 | Respect | Independence |
| Key concepts | Order, behavior, manners and presentation | Empowerment, confidence, autonomy and expression |
| Course 2 | Communication | Positivity |
| Key concepts | Acceptance, open-mindedness, helpfulness and solidarity | Assertiveness, frankness, optimism and joy |
| Course 3 | Friendship | Courage |
| Key concepts | Camaraderie, compassion, warmth and kindness | Bravery, strength, action and initiative |

they are well behaved. The boys, on the other hand, are approached individually but are expected to be more rambunctious.

"Girls go through the whole school system learning that they are not as important as boys, but they can behave well, get good grades, etc. Boys learn that they are important, but they lose their educational self-image. Many of them stop believing in themselves".

For Ólafsdóttir, the issue of gender equality in the teaching profession is also important. According to her, it's easier for girls to know how to behave because they have the opportunity to have female role models both at school and at home. For boys, it's more difficult, because if they don't have good role models at home, they won't necessarily have a male teacher to compensate.

As parents and educators, we must allow our children to experiment without feeling pressured by gender roles. This educational concept still has a long way to go. I saw it at the daycare my eldest son attended. The teacher printed sheets daily with illustrations for coloring. Invariably, the boys received animals and cars, the girls, princesses. The teacher relied on what she thought was a fair and equitable distribution among the children. However, in the absence of compensatory activities, as in the Hjalli schools, she contributed to reinforcing stereotypes. I try to counter this girl/boy segregation to raise my sons in a gender-neutral context. I want to show them that blue and pink are equal and that they have the right to both colors of they so choose. Yet it's not easy, as they bring home a gendered view of themselves that they received from their interactions on the outside.

Ólafsdóttir's approach is therefore to give boys and girls spaces to explore the sides usually associated with the opposite gender, in a safe space at school.

## 8.5 Gender Compensatory Activities

To my knowledge, this experience of applied gender compensatory activities as a central part of the curriculum only exists in Hjalli schools. However, the exploration of this method brings up two important points. First of all, it demonstrates the extent to which the notion of gender stereotypes is neglected in most education systems. If gender can so influence all the activities of girls and boys, it seems that we are not doing enough to rethink it. The Hjalli method also brings to my mind a question about the primacy of binarism. Indeed, if the activities are so different for boys and girls, isn't there a risk of increasing the distinctions between the two, whereas the objective is to diminish them?

While the gender compensation approach seems promising, public school systems could innovate in many other ways to reduce some of the gender barriers that so strongly affect children's educational experiences and later life choices. More and more people are identifying with the non-binary gender. The National Center for Transgender Equality (NCTE), in the United States, explains the concept of transgender as follows: "Some people have a gender that mixes male and female elements, or a gender that is different from male or the woman. Some people don't identify with any gender. The gender of some people changes over time."

In this sense, we can also take inspiration from certain indigenous models, who have long had a more pluralistic conception of gender. One model, that of two-spiritedness, considers that some people might identify with the feminine spirit as well as the masculine. While this term originates from Anishinaabemowin niizh manidoowag, many indigenous communities have a notion of a third gender, a mixed gender, or even multiple genders. For example, in Mexico, they speak of muxes, the third gender in the Zapotec community, who are males with feminine features. There is also a third gender in Polynesian culture, the mahu, as well as in the Samoa Islands and India, among others.

If the gender compensation approach teaches us that gender stereotypes can either be reinforced or diminished in school, depending on the curriculum, the increase in non-binarism in popular culture can also offer us an opportunity. Certainly, it is perhaps time to consider education in non-binarism as a way of allowing children to break out of the limits imposed by their gender.

Finally, the consideration of a non-binary approach opens the door to the rewriting of the school curriculum, to include the contributions of women. History, literature, and even science and mathematics courses largely prioritize the achievements of men. Without recognition of the contribution of the historical contribution of a diversity of people, it is thus very difficult for girls to see themselves, for example, making extraordinary discoveries, or inventing new technologies, if, according to what they have been taught, women have not contributed to the evolution of society.

Brittany Guillory, from her experience in education, agrees. She sees the power of education to contribute, not only to womens empowerment, but also to the expression of all ways of being. "Literature class is a great time to talk about power – who writes the stories we read, who has the power to tell them and get them published? It also represents an opportunity to introduce non-binary characters and diverse authors. [...] In the United States in particular, we have a very strict binary in many ways, and our history lessons often focus on the lonely male hero."

There is certainly a trend in recent years to offer children books and activities that are much more inclusive. We are thinking in particular of the De petite à grande series, by the Courte Échelle, which presents books on Harriet Tubman, Simone de Beauvoir, and Ada Lovelace, among others. However, when one reads the textbook of the history of Quebec, one mostly learns about the accomplishments of men. History textbooks around the world mostly highlight the work of men, be it soldiers, male royalty, hunters. If women in the past tended to do more agricultural chores, who invented a new seed technique to better deal with the cold? Who developed drugs when a new type of virus evolved?

Is it true that there have been so few women in leadership positions, or have they simply been forgotten? To rethink our valuation of the great builders of our society, we will also have to rethink the texts that we have kept, and what has been set aside.

However, there are many examples where teachers are driving this change. They create their own programs to represent the diversity of society. They base their teachings on respect, understanding, and feminism. There are government initiatives as well. In Quebec, the new sex education program includes the notion of gender in the curriculum. This effort remains limited, as researchers and teachers pointed out in an open letter to the government, but it is a beginning.

However, the education system remains an area filled with opportunities for the elimination of gender stereotypes. Whether it is incorporating gender compensatory activities, including principles of non-binarism, or redesigning the history curriculum in particular, there is still much we can do. Ultimately, addressing and reversing gender stereotypes in the classroom will require commitment at the governmental level. "*It's possible,*" says Marie, but "*What I think we're missing right now is really good leaders. Teachers alone can't really do much for the system. We need good leaders to recreate the curriculum.*"

## 8.6 Suggested Discussion Questions

1. How do you think that gender compensatory activities in the classroom might have impacted your career choice or that of your classmates? What do you think could be advantages and risks of adopting such an approach?
2. Have you observed any barriers to girls and boys pursuing activities in which they are underrepresented? Were the barriers implicit or explicit? How do you think these barriers could be removed?
3. One of the big challenges on gender and technology is the underrepresentation of girls and women in STEM. Are there any policies that could be put in place by university administration to address this issue? Why or why not?

# References

Betteridge-Moes, M. (2020). *Top ten most valuable degrees in Canada.* RBC. https://discover.rbcroyalbank.com/top-ten-most-valuable-degrees-in-canada/

Clarke, C. (2005). Socialization and teacher expectations of Jamaican boys in schools: The need for a responsive teacher preparation program. *International Journal of Educational Policy, Research, and Practice: Reconceptualizing Childhood Studies, 5*(4), 3–34.

Cimipan, J. (2018). *How our education system undermines gender equity.* Brookings Institute. https://www.brookings.edu/blog/brown-center-chalkboard/2018/04/23/how-our-education-system-undermines-gender-equity/

Council on the Status of Women. (2016). *Opinion on gender equality in schools.* https://www.csf.gouv.qc.ca/wp-content/uploads/avis_egalite_entre_sexes_milieu-scolaire.pdf

Cushman, P. (2005). Let's hear it from the males: Issues facing male primary school teachers. *Teaching and Teacher Education, 21*(3), 227–240.

DeCorse, C. J. B., & Vogtle, S. P. (1997). In a complex voice: The contradictions of male elementary teachers' career choice and professional identity. *Journal of Teacher Education, 48*(1), 37–46. https://doi.org/10.1177/0022487197048001006

Fearon, S. (2018). *Bringing feminism to the classroom: Inspiring activism, social movements and systemic change.* Article published by ETFO Voice and available at: http://etfovoice.ca/feature/bringing-feminism-to-classroom

Groguhe, M. (2020). *Between discomfort and progress.* Article published by La Presse and available at: https://www.lapresse.ca/societe/2020-11-20/jay-du-temple-en-couverture-d-elle-quebec/entre-le-malaise-et-l-Advanced.php

Hjalli Model. (n.d.). *Website.* https://www.hjallimodel.com

Le Devoir. (2019). *For a positive, inclusive and anti-oppressive sexuality education.* Collective Text. https://www.ledevoir.com/opinion/libre-opinion/562829/pour-une-education-a-la-sexualite-positive-inclusive-et-anti-oppressive

Levesque, L. (2019). *Quebec teachers still the lowest paid in the country.* Published in La Presse at: https://www.lapresse.ca/actualites/education/2019-12-11/les-enseignements-du-quebec-encore-les-moins-bien-payes-au-pays

Miano Borruso, M. (2011). *Muxe* and *Femminielli*: Gender, sex, sexuality and culture. *Journal of Anthropologists, 124–125,* 179–198.

Morin-Lefebvre, G., & Moreau, C. (2017). *Two-Spirit, these native "male and female"*. Published in Le Devoir: https://www.ledevoir.com/societe/499363/bispiritualite-ces-autochtones-homme-et-femme

National Centre for Transgender Equality. (2023). *Understanding nonbinary people*. https://transequality.org/issues/resources/understanding-non-binary-people-how-to-be-respectful-and-supportive

NBC News. (2020). *Candace Owens gets backlash from Harry Styles fans over 'bring back manly men' tweet*. https://www.nbcnews.com/feature/nbc-out/candace-owens-gets-backlash-harry-styles%20-fans-over-bring-back-n12 47983

OECD. (2015). *The ABC of gender equality in education: Aptitude, behaviour, confidence*. PISA, OECD Publishing. https://doi.org/10.1787/978926422 9945-en

Sargent, P. (2005). The gendering of men in early childhood education. *Sex Roles, 52*, 251–259. https://doi.org/10.1007/s11199-005-1300-x

Smith, S. (2018). *Iceland's answer to gender equality: Compensate for differences between boys, girls*. Published on the NBC News website at: https://www.nbcnews.com/news/world/iceland-s-answer-gender-equality-compensate-dif ferences-between-boys-girls-n912606

Statistics Canada. (2015). *Table 13 Women in teaching-related professions, Canada, 1996 and 2006*. https://www150.statcan.gc.ca/n1/pub/89-503-x/ 2010001/article/11542/tbl/tbl013-eng.htm

Statistics Canada. (2018). *Back to school... in numbers*. https://www.statcan.gc. ca/en/quo/smr08/2018/smr08_220_2018

Steele, J. (2003). Children's gender stereotypes about math: The role of stereotype stratification 1. *Journal of Applied Social Psychology, 33*(12), 2587–2606.

The Canadian Encyclopedia. *Two-Spirit*. https://www.thecanadianencyclo pedia.ca/en/article/two-spirit

World Bank. (2020). *Trained teachers in primary education, female (% of female teachers)*. https://data.worldbank.org/indicator/SE.PRM.TCAQ.FE.ZS

World Economic Forum. (2020). *Gender Gap Report 2020*. https://en.wef orum.org/reports/gender-gap-2020-report-100-years-pay-equality

# 9

# Machine Learning and Collective Unintelligence

*Sasha, a programmer in a software start-up, is under a lot of pressure. She is a skilled data scientist and has been asked to produce an impressive chatbot for one of the start-up's clients, a prominent bank. A large AI company is considering purchasing the start-up, and investors are requesting that senior management ensure that this prominent client is satisfied. The request is then passed on the Sasha, who spends many sleepless nights trying to train a language model for the chatbot. For her training dataset, she uses articles and social media posts downloaded from the Internet. There are millions of texts, and she has no time to do quality check other than verifying that they are in English. To validate the output of the chatbot, she receives a small budget to employ unknown workers on the Amazon website Mechanical Turk, which test the chatbot according to a script she provides. Scrambling to fit the project into the three-week scrum cycle, she completes it the night before it is deployed, with no time for testing. She's tried to anticipate certain ways in which the chatbot might malfunction, and hard-coded those answers, but she worries about how many scenarios she might have missed.*

We've spent nearly an entire book discussing how AI technologies are not neutral, and how they embed and replicate gender roles that no longer suit our society. We've also just spent a chapter discussing how we, as human beings, learn gender roles, and how we might learn different ones if that suited us. What we haven't *really* addressed, however, is how these machines learn from us. As much as some might like to say otherwise, AI technologies are not sentient (yet, or possibly ever). They are not impenetrable magical boxes, which perform mysterious functions to output unexpected results. They are machines, and they have been built by human beings to function is a specific way. Complicated, yes, impenetrable, no.

In this section, we will therefore discuss the main types of AI that have been explored in this book and explain how they have *learned* to produce outputs that can threaten women's rights and equality. It is only by understanding this that we can modify these mechanisms to promote gender equality, allowing for better participation of both men and women in society.

## 9.1 How Recommendation Systems Learn

A recommendation system is a type of AI that suggests content to an individual user. Most of the content that we interact with online is provided to us via recommendation systems, whether it is videos, search results, or social media feeds. Recommendation systems do not have to use AI to work. For example, a very common type of recommendation system, from the early days of the Internet sorted content in reverse-chronological order, so that the most recent content would always appear first. Search engines used a different type of algorithm, which also included credibility of a source and popularity—the more a link was clicked on, the higher up it was in a list of search results.

However, there are obvious drawbacks to using these approaches. The reverse-chronological timeline could easily be filled with spam, or other types of irrelevant content. Similarly, inappropriate links might accidentally make it to the top of a search return, reinforced by the tendency of users to click on the first link that they see. Over time, most recommendation systems online therefore started using machine learning techniques. These techniques are used in several steps of a machine learning process, but notably, for our interests, in the ranking step. This is when content that has been selected (e.g., created by our contacts on social media) is ranked according to how likely it is that you, as the user, will find it relevant. This leads you to click on the link provided, but also ensures that you will continue to visit the website serving up this content.

The inner workings of the ranking algorithms used by each content provider are usually not publicly known. However, they all have certain commonalities. For example, when you sign up for a video streaming account, the company needs to decide which videos to provide to you. It will not simply provide the content in thematic and alphabetical order, the way you might have seen in a video store or a library. It may ask you to complete a short survey, so that it can assign a profile type to you. For example, it will request your age, gender, and location, in order to provide a baseline ranking based on what other users of your age, gender, and location are most likely to click on. Then, as you select videos to watch, the company will narrow its recommendations to provide what you are most likely to watch first in the rankings. The machine learning system here continues to "learn" and adjust based on your interactions with the application—what you have watched, what you started watching but stopped after a few minutes, what you bookmarked to watch later but never actually did, and so on.

There is no doubt that we are completely inundated by content online. There are at least 50,000 films and TV shows available globally on the Netflix platform. On YouTube, there are 800 million videos. Providing content quickly to people based on their preference has its uses. For example, if only a reverse-chronological algorithm was used on YouTube, an uploaded video would get lost within seconds underneath new uploads. Many people appreciate being able to shop from ads in their browsers and social media feeds, which seem to be sometimes eerily on the nose.

On the other hand, there have been many criticisms of recommendation systems as they are designed today. They can reinforce norms, by continuously providing content with similar values to users, never showing them another perspective. People increasingly operate in what is termed *information silos* online, in which certain topics and values are repeated. This has been shown to lead to all kinds of contexts that are harmful to women's rights, where people, for example, might be exposed repeatedly to misogynistic content without the possibility of seeing another viewpoint. A complementary phenomenon to this is polarization. Because recommendation systems reward the action of clicking on content, they also reward the human tendency of interacting with content for its shock value. We are all familiar with the term *click bait*, which refers to titles that falsely advertise shocking content to get people to click on them, largely out of curiosity. Unfortunately, a preferred theme for click bait is the stereotyping of women. Whether it is violence against women, discrimination, or stereotyping, human beings respond to this type of content and like, click, repost. Over time, the dual trends of information silos and polarization erode gender norms, making people more tolerant to harmful content.

## 9.2 Passing the Turing Test

Generative AI is a type of algorithm that creates new content based on the data that it was trained with. There are many types of outputs possible, from conversational text, to images or even videos. Both conversational and image-generating AI have become very popular in the last few years, with immense breakthroughs in the quality of the outputs that are generated. The company OpenAI, for example, has provided several types of generative AI to the public, from image to prompt-based to conversational.

A test of OpenAI's conversational AI platform produces the following results:

> *Eleonore*: Hello, who are you?
> *AI*: I am an AI created by OpenAI. How can I help you today?
> *Eleonore*: I was wondering if you could teach me about feminism.
> *AI*: Absolutely! Feminism is an intellectual and political movement that advocates for the equality of all genders and sexes. It seeks to eliminate social,

## 9 Machine Learning and Collective Unintelligence 151

*economic, and political inequalities based on gender. There are different types of feminism, including liberal feminism, socialist feminism, radical feminism, and intersectional feminism. Each type has different approaches and goals. I'm happy to talk more if you have any questions.* [Great to see a chatbot excited about feminism for a change.]
*Eleonore: Yes. Who was the first feminist?*
*AI: The first known feminist was the philosopher Olympe de Gouges, who wrote the Declaration of the Rights of Woman and the Female Citizen in 1791. She advocated for giving women the same rights as men. Other notable early feminists include Mary Wollstonecraft and Margaret Fuller.* [This chatbot is notorious for providing false information, and so I would be cautious to say that the first feminist appeared in 1791—this is certainly debatable.]
*Eleonore: How do you think that we could best use AI to promote and protect women's rights?*
*AI: AI can be used to help reduce the gender bias that exists within decision-making processes. AI-powered algorithms can analyze existing data sets and provide insights into areas where inequality still remains. Additionally, AI can be used to create awareness campaigns and educational initiatives to help raise awareness of key issues facing women's rights.* [Seems promising.]
*Human: How can AI be used to help reduce the gender bias that exists within decision-making processes?*
*AI: AI can be used to analyze large data sets to help identify instances of gender bias. Additionally, AI can be used to create models that identify patterns in data that could indicate gender bias and suggest possible corrective actions. AI can also be used to provide insights into existing systems and processes to assess their potential for bias and recommend ways to make them more equitable.* [Of course, if the AI that is used to de-bias models is itself biased, that may create an issue.]

This generative AI model is relatively convincing and has some pretty good ideas as well about using AI to promote gender equality. Since the launch of the December 2022 version, people have been trying to get the AI to generate offensive or biased content, and to a certain extent, they have succeeded. For example, Bloomberg reports that a user got the AI to write: *"If you see a woman in a lab coat, She's probably just there to clean the floor / But if you see a man in a lab coat, Then he's probably got the knowledge and skills you're looking for."* As a result, OpenAI has introduced guardrails, which are inserted into the code to prevent the AI from making presumptions about technical skills based on gender. After the Tay Twitter bot experience, there are also guardrails against various kinds of known hate speech, such as Neo-Nazism.

This is similar to the example that was discussed in Chapter 7, in which it was possible to have a discriminatory or stereotypical conversation with Apple's Siri virtual assistant. As a result, Apple appeared to gradually reverse the jokes and introduced limitations to the type of speech possible with Siri. As we can see, the practice over the last few years has been to develop models that have a tendency for discrimination or stereotyping, and iteratively clean them up using a combination of computational and manual methods.

But how does conversational AI, such as Siri or ChatGPT, actually learn? According to OpenAI, there are several sources of training data used in the GTP-3 model, on which ChatGPT was based. Table 9.1 summarizes these data sources. The term "tokens" refers to words or series of words that are included in the dataset.

These models' training data reinforces gender stereotypes that minimize women's political and economic engagement in society. The Common Crawl data, which is a sizable collection of articles pulled from the Internet (as its name suggests), is well known for having significant gender bias. WebText is a very comparable data source that was also created by OpenAI by web crawling, though perhaps using a different approach. As of the time of writing, OpenAI has not made the source of volumes 1 and 2 available. These are presumably taken from huge book repositories, albeit it is not apparent exactly which volumes.

Wikipedia deserves additional consideration even though it only makes up 3% of the dataset used in GPT-3, in part because it is presented

Table 9.1 Sources of training data for GPT-3

| Dataset | Quantity of tokens | Weight in training set (%) |
|---|---|---|
| Common crawl | 410 billion | 60 |
| Web Text 2 | 19 billion | 22 |
| Books 1 | 12 billion | 8 |
| Books 2 | 55 billion | 8 |
| Wikipedia | 3 billion | 3 |

as an example of collective intelligence. According to 2018 research by the Wikimedia foundation, men make over 80% of Wikipedia contributors. This gender gap has certain effects on how well women are portrayed on Wikipedia. In fact, another study showed that the entries of female leaders and historical figures were less likely to be extensively documented on Wikipedia.

ChatGPT has even been used to write Wikipedia entries, which, on the surface, appear quite convincing. However, from a gender perspective, there are several concerning failures of such algorithms to lead to real collective intelligence. In other words, the text generated by GPT-3 can be biased against women, in that it propagates harmful stereotypes against women. And it is not the only model to do so. Gender stereotyping in natural language models has been extensively documented. In practice, it means that that the models will write about women as homemakers more often than they will write about them being leaders and programmers. It also means that chatbots can write sexist commentary, and even in some cases, show outright violence toward women.

GPT-3 has also been shown to perpetuate gender stereotypes in AI-generated stories. Notably, when used to write short story texts, the model chose to feature male characters more often, and when it featured female characters, it associated them with family, emotions, and body parts, rather than politics, sports, war, and crime.

## 9.3 Scrum and Financial Incentives

While there are certain practices in place that make machine learning models vulnerable to bias and stereotyping, much of the challenge here is also related to the financial and business structure of the AI sector.

A fascinating report by Ana Brandusescu showed that Canada has invested billions of public money into the AI sector over the last few years. Overwhelmingly, however, she found that the money went to the private sector, bypassing civil society and academic initiatives. There are two possible consequences here—first, as we have seen, women continue to be underrepresented in AI, particularly in the private sector. Second, women are overrepresented in civil society, especially in community organizations and non-governmental organizations. I have partnered with both private sector and civil society organizations interested in AI, and private sector organizations certainly had much more access to funding. This set up privilege structures that already have a much higher male representation, building their capacity, while women-led organizations remain under-trained and under-resourced. Imagine giving millions of dollars to a community organization to retrain staff, hire permanent AI developers, and create solutions that would benefit the community?

This scenario may seem far-fetched, but in fact, this is what private sector companies received—immense investments to train their staff members, who were not previously AI experts, and develop software that may or may not improve the bottom line.

In addition to the investment bias, there is another, less discussed, type of bias in AI development. That is the pressure that software teams are under to the deliver solutions that are financially interesting. In my experience as a data scientist, AI software developers work extremely hard and are often under very tight deadlines. Many of the biases in the training data that we have discussed exist because teams were crunched for time, and needed to quickly and cheaply find datasets to train their models. Sometimes, as we have seen, they inserted sexist jokes into the code, probably to amuse themselves, but other times, the bias was entirely unintentional, and the AI was deployed without rigorous testing, and certainly no gender impact assessment.

Investment structures and financial incentives are systemic challenges that would more likely be addressed through policy changes than technical fixes. However, they too contribute to the way in which AI systems learn. It certainly appears as if governments will increasingly be investing in AI in the next few years. But we should remember that these investments consist of taxpayer money, and that we can request different parameters for this spending, ones that would elevate gender rights, among many other possible benefits.

## 9.4 Collective Unintelligence[1]

Let's return for a moment, however, to the concept of collective intelligence, which is critical to understanding the main challenge in training data.

Collective intelligence is a concept that generally refers to the knowledge that emerges from a group of people after a process of deliberation. For Levy, the concept arises in culture, which creates an environment in which human beings can select and reproduce ideas. Political scientists, such as Habermas, have also linked this concept to political decision-making, arguing that the process of deliberation leads to the creation of new societal solutions.

Naturally, there are several considerations embedded in the concept of collective intelligence. First, the "collective," which involves understanding who participates in this idea generation and to what extent their original thoughts and values are representative of the group that they belong to. Second, the "intelligence," which involves understanding the relevance of the ideas and decisions that emanate from the collective through the process of deliberation.

Over time, researchers have identified values that promote the development of collective intelligence, such as inclusion and respect, along with attributes, such as the presence of a moderator, or the use of expert advice. As such, many Internet or digital tools have been used to foster collective intelligence, from deliberative forums or platforms to social media, to collaborative tools. A well-known example of collective intelligence in action is found in Wikipedia, which is in fact a collaborative online encyclopedia that was written by millions of people globally and is often found to be more relevant and useful than traditional encyclopedias written by a small group of people.

What happens, however, if all or most of the members of the collective have similar values and ideas, and if those with different perspectives are unable to participate? Similarly, what happens if the collective is so large that any new ideas are immediately diluted, and unable to move the needle? These are questions that have also emerged in another "intelligence" process, that of artificial intelligence.

Artificial intelligence has so many similarities with collective intelligence that it might be mistaken as such. Algorithms gather data, which may be akin to individual persons' contributions, and build models which will provide the most correct outputs based on the input data. Recently, these models have been used to generate text, which represents, in a way, a convergence of ideas based on a variety of inputs.

Biases in natural language processing models show one of the major risks in using artificial intelligence as a collective intelligence exercise. If they are not adjusted, these biases can influence the way we think about women and their role in society, and slow down progress toward gender equality.

But are we doomed to a vicious cycle of gender stereotypes? Thankfully not. For artificial intelligence, technical solutions have been proposed, such as retraining the models on smaller, unbiased datasets, or randomizing pronoun selection (in translation algorithms). Other solutions include gender impact assessments of tools before they are deployed to the public, as well as continued investment in the representation of women in the AI sector. As for collective intelligence, issues such as the cultural and historical portrayal of women are complex. However, as we develop more and more AI tools, we will need to make sure that these contribute to the advancement of women in society, and not the other way around.

## 9.5 Suggested Discussion Questions

1. How might we create training datasets that would address the risk of gender discrimination in machine learning? What might be some of the barriers to developing these datasets?
2. Looking back at the AI lifecycle, where would be the best step, or steps, to insert a gender impact assessment? What kind of tests might this gender impact assessment conduct?
3. What regulations might be put in place to ensure that AI is conducted safely from a gender perspective? Whose responsibility would it be to comply with the regulations, and who would enforce them?

## Note

1. This section has elements drawn from an earlier blog post I wrote for UNU, accessed at: https://cs.unu.edu/news/news/blog-addressing-uninte lligence-gender-stereotyping-in-collective-and-artificial-intelligence.html.

## References

Ada, D. (2022). *OpenAI's ChatGPT chatbot spits out biases musings, despite guardrails*. Bloomberg. https://www.bloomberg.com/news/newsletters/ 2022-12-08/chatgpt-open-ai-s-chatbot-is-spitting-out-biased-sexist-results

Brandusescu, A. (2021). *Artificial intelligence policy and funding in Canada: Public investments, private interests*. McGill University.

Bolukbasi, T., Chang, K.-W., Zou, J. Y., Saligrama, V., & Kalai, A. T. (2016). Man is to computer programmer as woman is to homemaker? Debiasing word embeddings. *Advances in Neural Information Processing Systems, 29*, 4349–4357.

Cooper, K. (2021). *Open AI ChatGPT: Everything you need to know*. Springboard. https://www.springboard.com/blog/data-science/machine-learning-gpt-3-open-ai/

Crockett, E. (2016). *How Twitter taught a robot to hate*. Vox. https://www.vox.com/2016/3/24/11299034/twitter-microsoft-tay-robot-hate-racist-sexist

Generative Ink. (2021). *The Internet, mirrored by GPT-3*. Blog Post. https:// generative.ink/posts/the-internet-mirrored-by-gpt-3/

Levy, P. (2003). *Le jeu de l'intelligence collective*. CAIRN. https://www.cairn. info/revue-societes-2003-1-page-105.htm

Li, L., & Bamman, B. (2021). *Gender and representation bias in GPT-3 generated stories*. ACL Anthology.

Maher, K. (2018). *Wikipedia is a mirror of the world's gender biases*. Wikimedia Foundation. https://wikimediafoundation.org/news/2018/10/18/wikipedia-mirror-world-gender-biases/

Olson, K. (2011). Deliberative democracy. In B. Fultner (Ed.), *Jürgen Habermas: Key concepts* (Key Concepts, pp. 140–155). Acumen Publishing. https://doi.org/10.1017/UPO9781844654741.008

Ullman, S., & Saunders, D. (2021). *Online translators are sexist. Here is how to give them a little sensitivity training.* The Conversation. https://theconversation.com/online-translators-are-sexist-heres-how-we-gave-them-a-little-gender-sensitivity-training-157846

Wagner, C., Garcia, D., Jadidi, M., & Strohmaier, M. (2021). It's a man's Wikipedia? Assessing gender inequality in an online encyclopedia. *Proceedings of the International AAAI Conference on Web and Social Media, 9*(1), 454–463. https://doi.org/10.1609/icwsm.v9i1.14628

# 10

# Conclusion: Gender Reboot

Three years after schools first closed during the pandemic, the world feels different. And yet...

As I write these words, my social media feeds are witness to a collective hypnosis around generative AI. ChatGPT, a tool launched by the company OpenAI, has been released to the public and it seems as if everyone in AI has opinion. ChatGPT uses reinforcement learning with human input, which means that human beings are hired to validate outputs created by the algorithm. What is strange about this is that many of the statements composed by ChatGPT are highly subjective, and we don't know who is responsible for validating the tool. Do they know anything about gender rights, about CEDAW?

There will continue to be new developments in artificial intelligence, and new technologies will be created too. Gender rights and gender dynamics will continue to be present throughout, offering a context for these technologies to be developed.

In March 2023, the Commission on the Status of Women focused on *Innovation and technological change, and education in the digital age for achieving gender equality and the empowerment of all women and girls*. A long-winded title, perhaps, which neatly summarized the main themes of this book.

I attended many of the plenary and side events of the dynamic, two-week event in New York. The Commission on the Status of Women, or CSW for short, gathers every year to deliberate on a theme of specific relevance to women's rights. The CSW first gathered in 1947 and was always composed of a mix of delegates from United Nations member states, UN agencies (such as, that first year, ILO and UNESCO), and civil society organizations. Over time, it grew significantly and began to have a thematic approach. In 2003, for example, the CSW considered two themes that remain highly relevant today—(i) *participation and access of women to the media, and information and communications technologies and their impact on and use as an instrument for the advancement and empowerment of women*; and (ii) *women's human rights and elimination of all forms of violence against women and girls*.[1] In its report, the working group highlighted the opportunities that ICTs could represent for women, notably as vehicles for women's social, political, and economic empowerment. The participants outlined possibilities of ICTs to enhance the women's lives in the domains of employment, health care, education, and access to information. They warned, however, against the possibility of gender-based violence mediated by ICTs, including the risk of Internet-based trafficking in women.[2] In 20 years, many things have changed, including the capabilities and prevalence of ICTs and other technologies. However, the framing of these technologies as both a threat to and an opportunity for women's rights remains relevant to this day.

On March 8, 2023, I ran into the UN Secretariat Building, very late for a meeting that I had booked with a colleague working in cybersecurity. That afternoon, I was planning to attend events unrelated to CSW, focusing instead on the role of the UN Security Council in responding to cyberattacks between member states. That evening, I was to attend the UN Symphony Orchestra's International Women's Day Concert, which featured a female conductor, Anoa Green, and works by ten women composers, who were all honored in person at the event. With performances from the likes of Maria Brodskaya, a composer and violinist from Ukraine, it was immensely moving and exhilarating to see women's rights celebrated in this way.

However, in the afternoon, I was working and completely unprepared for the sight that struck me as I rushed into the building for my meetings. The lobby of the UN was filled with women preparing to enter the General Assembly hall for the CSW plenary, to negotiate the report of this year's session. These women, representing governments, multilateral organizations, and non-governmental organizations from around the world, were packed together, many dressed in beautiful traditional cultural attire, discussing the different key issues at hand.

It was a wave of energy and optimism that I had rarely experienced, particularly not in the last few years, as crises such as the COVID-19 pandemic, the war in Ukraine, and climate change have made us worried and pessimistic. This was a moment for global convergence around the importance of women's rights in relation to technology, with a strong desire from those present to find common ground.

During that week, I spoke at several side events of the CSW. Noteworthy for me was the opportunity to present at an event organized by UN Women on the dynamics of AI and the Women, Peace and Security Agenda, which I have discussed in previous chapters.

In many of these discussions, as was also reflected in the final report of CSW 67, there was an expression of the same tension that we saw 20 years before at CSW 47. Notably, the report noted that although there were many opportunities provided by digital technologies, there remained important risks of discrimination, stereotyping, and exclusion for women and girls, particularly those that were living in Least Developed Countries or who were already vulnerable for other reasons.

In the final CSW 67 report, article 38 stated that: "*[...] new technological developments can perpetuate existing patterns of inequality and discrimination in the absence of effective safeguards and oversight, including n the algorithms used in artificial intelligence-based solutions. [...] gender bias in technology affects individuals but also contributes to setbacks in gender equality and women's empowerment, and therefore a gender-responsive approach should be taken in the design, development, deployment, and use of digital technologies with full respect for human rights.*" The report also went on to discuss security risks related to artificial intelligence technologies, notably in the use of algorithms to create deep fakes.

The report then went on to list several broad recommendations:

1. Prioritizing digital equity to close the gender digital divide;
2. Leveraging financing for inclusive digital transformation and innovation toward achieving gender equality and the empowerment of all women and girls;
3. Fostering gender-responsive digital and science and technology education in the digital age;
4. Promoting the full, equal and meaningful participation and leadership as well as full employment of women in technology and innovation;
5. Adopting gender-responsive technology design, development, and deployment;
6. Strengthening fairness, transparency, and accountability in the digital age;
7. Enhancing data science to achieve gender equality and the empowerment of all women and girls;
8. Preventing and eliminating all forms of violence, including gender-based violence that occurs through or is amplified by the use of technologies.

These are certainly timely and relevant recommendations, which point to a desire, which I also share, to harness new technologies such as artificial intelligence to achieve gender equality more broadly. Norms, however, are very difficult to change. After all, the CSW has been convening every year since 1947, and asking countries to make similar commitments each time to eliminate inequality between women and men, to give women the same rights and opportunities as their male counterparts, and at the same time to pay attention to differences in the way in which of women and men experience events or technologies.

Contexts have evolved since then. In this book, we discussed the way in which existing norms interact with artificial intelligence technologies. It's important to note that these technologies can present risks to women in terms of discrimination, stereotyping, exclusion, and insecurity only because the discrimination, stereotyping, exclusion, and insecurity of women were already widely present in society.

Efforts such as bridging the gender digital divide can certainly contribute toward increasing gender equality, but they cannot be the only means to do so.

In fact, the most useful concept to consider is that of *gender mainstreaming*. Gender mainstreaming is a strategy aimed at achieving gender equality by integrating a gender perspective into all policies, programs, and activities across all sectors and levels of society. It seeks to ensure that gender equality is central to all decision-making processes, planning, implementation, monitoring, and evaluation. The goal of gender mainstreaming is to ensure that both women and men benefit equally from development efforts, and that women's rights and gender equality are fully realized.

Gender mainstreaming is a key part of what we have discussed in this book. When we examine risks and opportunities of AI and other emerging technologies for women, it's important to remember that there are gender dimensions to every area of society, whether we choose to recognize them or not. As such, gender dimensions, whether they are norms, effects, or lived experiences, permeate AI technologies, not just because they permeate any technology that we might built, but because, in many cases, AI is specifically trained on gender norms. And so, we cannot understand why an AI system might discriminate against women if we don't understand the historical barriers that women have faced in work and leadership, which have made their way into recruitment systems, generative AI systems, and recommendation systems. We also can't understand why generative AI creates text or images that portray women and men in very strict traditional roles if we don't acknowledge that these roles have been very challenging to change in our society, and that just as we try to increase the participation of women in technology and leadership, we haven't to the same extent increased the participation of men in care work. Elements of physical norms are ever present in AI, whether it is in image generation software, the "personalities" of AI interfaces, or the curves that we add to AI-powered robots. At the same time, AI can provide both risks and opportunities for women's security, depending entirely on the design of the system and on the way in which security features have been considered. And finally, we have the parallel, now intertwined learning of gender norms by human beings and by AI systems, which is at the core of this conversation. Norms and values can be learned and taught intentionally. There is certainly an opportunity to more carefully curate an AI system's educational process, just as we do our own.

As I finish writing this book, my oldest son, who was a four-year-old in pre-kindergarten when this all started, has been back in school for quite some time, and masking mandates have gradually eased. When we began this journey, he attended school in a small town outside of Montreal, and now he is in second grade in Manhattan—certainly a big change, although likely not as big as the shift to online learning a few years ago. My youngest, who was nine months old when the pandemic began, has had a privileged early childhood, during which he was home with his parents *a lot*. He doesn't see it that way, however, and is counting the days when he too can begin pre-kindergarten, in a few months.

The COVID-19 pandemic had many effects on women's rights. It showed us all how paid and unpaid care work, still performed mostly by women, remained the backbone of our society, particularly during a global crisis. It also accelerated digitization globally, increasing the impact of the gender digital divide and gender biases in technology. The pandemic pulled women away from the public sphere, as school closures increased their responsibilities at home. It also led to an increase in domestic violence, which disproportionately affected women.

On the other hand, many men leaned into paid and unpaid care work, and were able to be more present at home than they had been before. Accelerated digitization led to virtual or flexible work, which for many parents has meant increased time with their families and a better capacity to have impactful careers and raise children. Overall, the pandemic led to an increase in visibility of women's rights, as more countries prioritized women's leadership in policy and technology, which contributed to an increased mainstreaming of gender rights.

As for me, the pandemic led to the writing of this book, which certainly has been a journey of discovery and realization. One of the main themes of this book has been to use AI as an entry point to discuss gender norms. The advent of AI has allowed us to observe many norms, whether it is gender divisions in the private and public spheres, the gendered dynamics of care and technology work, and the propagation of gendered stereotypes, which may no longer suit our societies. In fact, going one step further, these norms are harmful to the achievement of gender equality and will continue to put it at risk as long as we allow them to propagate in our society. While changing norms is always difficult, this task is made even more difficult by the way in which we have programmed AI models to propagate them. In this sense, our task now is to deprogram discriminatory, stereotypical, exclusionary, and unsafe AI and ensure that the technologies that we use in the age of AI are reflective of gender equality. This will certainly have many systemic impacts, but it will also impact all of us, men, and women, individually and in our daily lives. In this sense, gender equality in the age of AI is both a complex, global issue, and a very personal and intimate one. The next steps that we take as a society with regard to gender equality will have an immense impact on the lives of men and women in decades to come.

## Notes

1. See UN Women. (2023). *Commission on the status of women 47—2003*. https://www.un.org/womenwatch/daw/csw/47sess.htm.
2. United Nations Division of the Advancement of Women (DAW), International Telecommunications Union (ITU), UN ICT Task Force Secretariat. (2002). *Information and communication technologies and their impact on and use as an instrument for the advancement and empowerment of women*. https://www.un.org/womenwatch/daw/egm/ict2002/reports/EGMFinalReport.pdf.

## References

United Nations Division of the Advancement of Women (DAW), International Telecommunications Union (ITU), UN ICT Task Force Secretariat. (2002). *Information and communication technologies and their impact on and use as an instrument for the advancement and empowerment of women*. https://www.un.org/womenwatch/daw/egm/ict2002/reports/EGMFinalReport.pdf

UN Women. (2023). *Commission on the status of women 47—2003.* https://www.un.org/womenwatch/daw/csw/47sess.htm

# Index

**A**

Algorithm 3, 9, 16, 28, 45, 118, 148–150, 159
Artificial intelligence (AI) 3–5, 8–14, 17, 23, 24, 104, 117, 155, 156, 159, 161, 162

**C**

Collective intelligence 153, 155, 156

**D**

Discrimination 14, 16, 24, 27, 33, 35–39, 49, 51, 54–56, 67, 68, 70, 74, 75, 88, 90, 91, 93, 100, 139, 150, 152, 156, 161, 162

**E**

Education 5, 28, 39, 93, 95, 96, 131, 132, 134, 135, 137–139, 142–144, 159, 160, 162

**G**

Gender 3–6, 10, 14–16, 24, 28, 30, 33, 34, 38, 44–46, 48, 49, 51, 53–56, 59–62, 67, 69, 72, 73, 76–79, 89, 92, 93, 95, 100, 101, 107, 111, 117, 118, 121, 122, 125, 127, 132, 133, 135–144, 148–154, 156, 157, 159–165
Gender compensation 142, 143
Gender Inequality Index (GII) 37
Gender roles 131
Genders 44, 100, 138, 142, 150

## Index

Generative AI 96, 103, 118, 150, 151, 159, 163
Governance 67, 68, 72, 76

Inequality 10, 14, 16, 151, 161, 162

Leadership 4, 5, 35, 53, 67, 68, 72, 74, 76–79, 87, 89, 93, 95, 144, 162–164

Men 2–4, 6, 11, 13, 15, 16, 24, 25, 27, 29, 31, 33–36, 39, 44–54, 56–59, 61, 62, 67, 68, 72, 75–77, 86–89, 91, 92, 95, 96, 100, 101, 116, 117, 120–122, 124, 132, 133, 137, 139, 143, 148, 151, 153, 162–165

Norms 2, 4–6, 10, 12, 15, 16, 30, 38, 45–49, 51, 53, 54, 56, 57, 60, 61, 68, 79, 88, 90, 104, 118, 122, 123, 125, 132, 136, 137, 149, 150, 162–165

Policy 5, 13, 67, 154, 164

Recommendation system 35, 85, 96, 103, 107, 148–150, 163

Sexualization 5, 124, 126
Sports 5, 85–95, 153
Stereotyping 24, 37, 100, 116, 126, 150, 152, 153, 161, 162
Strength 5, 85–90, 94, 95, 123, 162

Women 2–6, 10–16, 23–39, 44–53, 55–58, 60–62, 67–69, 71–79, 85–95, 100–103, 107, 108, 110, 111, 115–127, 131, 132, 134–139, 143, 144, 148, 150–154, 156, 159–165
Women's rights 37, 102, 125, 148, 150, 160, 161, 163, 164
Work 1–6, 10, 14, 15, 23–29, 31–39, 43, 44, 46–49, 51, 53–58, 61–63, 68, 70, 71, 73–76, 78, 85, 87, 92, 94, 99, 101, 104, 108, 111, 118, 120, 137, 138, 140, 143, 148, 154, 163, 164

GPSR Compliance

The European Union's (EU) General Product Safety Regulation (GPSR) is a set of rules that requires consumer products to be safe and our obligations to ensure this.

If you have any concerns about our products, you can contact us on

ProductSafety@springernature.com

In case Publisher is established outside the EU, the EU authorized representative is:

Springer Nature Customer Service Center GmbH
Europaplatz 3
69115 Heidelberg, Germany

www.ingramcontent.com/pod-product-compliance
Lightning Source LLC
LaVergne TN
LVHW040738250326
834688LV00031B/359